The Jaguar Leg
of Luxury, Perfo
and Innovation

Etienne Psaila

The Jaguar Legacy: A Tale of Luxury, Performance, and Innovation

Copyright © 2025 by Etienne Psaila. All rights reserved.

First Edition: **February 2025**

No part of this publication may be reproduced, distributed, or transmitted in any form or by any means, including photocopying, recording, or other electronic or mechanical methods, without the prior written permission of the publisher, except in the case of brief quotations embodied in critical reviews and certain other non-commercial uses permitted by copyright law.

ISBN: 978-1-923432-25-3

Table of Contents

1. Introduction: The Birth of Jaguar – A Brand of Distinction
2. The Early Years: From Swallow Sidecar Company to Jaguar Cars
3. Jaguar's Evolution: Key Models and Innovations
4. Jaguar's Pioneering Role in British Automotive Design
5. The Jaguar XJ Series: A Legacy of Luxury Sedans
6. The Birth of the E-Type: The Icon of the 1960s
7. Jaguar's Racing Heritage: From Le Mans to Formula 1
8. Jaguar and the British Motor Industry: A Pillar of Excellence
9. The Changing Landscape: Mergers, Acquisitions, and Ownership
10. Design and Innovation: The Jaguar Approach to Crafting Cars
11. Jaguar and the American Market: A Tale of Love and Challenges
12. The XJ220: The Supercar That Almost Was
13. The Revival of Jaguar: The 21st Century and New Challenges
14. Jaguar's Contribution to Electric Vehicles: A New Era of Mobility
15. The Role of Jaguar in Luxury and Performance Car Culture

16. Jaguar's Global Expansion: How the Brand Conquered the World
17. Jaguar and British Luxury: The Intersection of Performance and Prestige
18. Technological Innovations: From the Monocoque to All-Wheel Drive
19. The Future of Jaguar: Electric Vehicles, Autonomous Cars, and the New Mobility Era
20. The Enduring Appeal of Jaguar: Why the Brand Remains a Symbol of Excellence
21. Jaguar in Popular Culture: A Symbol of Glamour and Speed
22. The Future of Jaguar: Challenges and Opportunities in the Electric Age
23. Conclusion: The Jaguar Legacy – A Brand That Transcends Time

- Appendix A: Key Models and Milestones
- Appendix B: Jaguar's Impact on Automotive Design
- Appendix C: Jaguar's Motorsports Legacy

Chapter 1: Introduction: The Birth of Jaguar – A Brand of Distinction

Jaguar, a name synonymous with elegance, performance, and innovation, stands as one of the most celebrated brands in automotive history. The story of Jaguar is not just the tale of a car manufacturer, but of a symbol of British engineering and design excellence. From its humble beginnings to becoming a globally recognized luxury automobile maker, Jaguar has captivated car enthusiasts and collectors worldwide.

Born from a rich heritage of craftsmanship and a desire to challenge the boundaries of design and performance, Jaguar has built a reputation for creating cars that aren't just modes of transport, but works of art. Whether it's the sleek and powerful E-Type or the sophisticated XJ, each model reflects the philosophy of blending beauty with power, luxury with performance, and technology with craftsmanship. Jaguar's cars have often been more than just vehicles; they have become icons of style, status, and the quintessential British driving experience.

This book takes you on a journey through the history of Jaguar, tracing its origins, breakthroughs, and its rise to prominence as a leader in the automotive world. The evolution of Jaguar is a fascinating tale of ambition, innovation, and the pursuit of perfection. From its early years as the Swallow Sidecar Company to its transformation into the Jaguar Cars we know today, the brand has always been about pushing the limits of what is possible in automotive design.

As we delve into the history of Jaguar, we will explore not just the cars, but the people behind the brand, the challenges it faced, and the profound impact it had on the automotive industry. We'll look at its role in British motoring heritage and its ability to captivate global markets. From the factory floor to the racetrack, Jaguar's legacy is a story of resilience and ambition, qualities that have kept it at the forefront of the luxury car market for decades.

Chapter 2: The Early Years: From Swallow Sidecar Company to Jaguar Cars

Jaguar's journey begins not with a car, but with a sidecar. It was in 1922, in the small town of Blackpool, that two entrepreneurs, William Lyons and William Walmsley, set up the Swallow Sidecar Company. The pair's initial focus was on creating high-quality sidecars for motorcycles, a booming industry at the time. The business quickly gained a reputation for fine craftsmanship, and soon they began to experiment with car bodies, which would set the stage for the company's future.

Swallow Sidecar's first car project was the 1927 Swallow Sidecar Company-built body for the Austin Seven, a popular small car in Britain. Lyons and Walmsley saw the potential for creating attractive, stylish car bodies for mass-produced vehicles, and the demand for these bespoke bodies soon grew. By the early 1930s, the company had expanded its operations, moving into full-scale automobile production. The name "Swallow" was no longer adequate to reflect the aspirations of the company, and in 1935, it officially changed its name to Jaguar Cars Ltd.

The 1930s marked a defining moment in Jaguar's history. Lyons, driven by a desire to create cars that offered more than just style, began to focus on performance and engineering. This ambition led to the development of the 1935 SS Jaguar 2.5-litre, the first car to feature the Jaguar name. The car was a masterpiece of design, with a sleek, elegant body that became a hallmark of Jaguar's future creations. It was more than just a pretty face; the SS Jaguar was powerful and fast, capable of competing with some of the best in the world.

However, the 1930s were also a time of challenges. As Europe slid toward World War II, the automotive industry faced difficulties, and Jaguar was no exception. The war effort brought a halt to civilian car production, and Jaguar had to adapt to the changing landscape. During this time, the company became involved in the production of military vehicles, which kept its workforce employed and its factories running.

In the aftermath of the war, Jaguar, like many other car manufacturers, was faced with the challenge of rebuilding and adapting to a peacetime economy. But Jaguar had something that few other manufacturers could boast – a strong brand identity, a reputation for quality, and an understanding of the emerging demand for affordable yet luxurious cars. The SS Jaguar, and later the introduction of the Jaguar XK120 in 1948, would cement Jaguar's place in the post-war automotive renaissance.

The XK120 was a turning point for Jaguar. Not only was it one of the most beautiful cars ever made, but it also offered exceptional performance. It was a true sports car, with a top speed that made it one of the fastest cars in the world at the time. Its success in motorsports and its exposure on racetracks around the globe would help build Jaguar's reputation as a performance brand. The XK120 also showcased Lyons' ability to marry engineering prowess with stunning design, a theme that would continue throughout Jaguar's history.

From the early days as a sidecar company to the creation of the first Jaguar-branded car, the journey to becoming a world-renowned manufacturer was marked by strategic decisions, hard work, and an unrelenting focus on quality. It was during these formative years that Jaguar established the foundations upon which its future success would be built.

As Jaguar moved into the 1950s, it faced the challenge of establishing itself in the increasingly competitive global automotive market. But with models like the XK120 and the introduction of the Jaguar Mark VII saloon, the brand was poised to not only survive but to thrive. These early years were pivotal, setting the tone for Jaguar's future as a symbol of British automotive excellence, combining sleek design, engineering innovation, and a commitment to performance.

Chapter 3: Jaguar's Evolution: Key Models and Innovations

Jaguar's journey from a fledgling car maker to an internationally recognized icon is defined by its relentless pursuit of innovation and its ability to create cars that are both technologically advanced and visually stunning. Throughout its history, Jaguar has produced several key models that not only define the brand but also leave a lasting impact on the automotive world. These models, through their design and performance, have set new standards for luxury, speed, and craftsmanship, each representing a chapter in Jaguar's evolution.

The XK120: The Birth of a Legend

When the Jaguar XK120 was launched in 1948, it marked the beginning of Jaguar's transformation into a performance brand. Designed by William Lyons and his team, the XK120 was a two-seater sports car that combined elegance with thrilling performance. Powered by a 3.4-liter inline-six engine, it was capable of reaching speeds over 120 mph, which was groundbreaking at the time. The XK120's sleek, curvaceous body and powerful performance helped establish Jaguar's reputation for creating beautiful cars that were also highly capable.

The XK120 wasn't just a success in showrooms; it dominated motorsports as well. Jaguar's victories at prestigious events like the Mille Miglia and Le Mans put the brand on the global map, showcasing its prowess both on the track and on the road. The XK120's impact cannot be overstated; it laid the groundwork for Jaguar's future as a performance-oriented brand while remaining true to its commitment to luxury and design.

The Mark VII: Jaguar's First Full-Sized Saloon

In addition to creating sports cars, Jaguar also sought to develop saloons that could embody its commitment to performance and luxury. The Jaguar Mark VII, introduced in 1950, was the company's first full-sized luxury saloon, and it further solidified Jaguar's position as a manufacturer of high-quality vehicles. The Mark VII featured a refined design, a spacious interior, and the same powerful engines that had made Jaguar's sports cars famous.

The Mark VII was a success, offering performance that rivaled some of the best sports cars of the time. It was powered by Jaguar's 3.4-liter XK engine, which provided more than enough power to make the Mark VII an exciting drive. Its success in both the consumer market and on the racetrack further reinforced Jaguar's growing reputation for producing cars that were fast, stylish, and luxurious.

The E-Type: The Icon of the 1960s

In 1961, Jaguar unveiled the E-Type, a car that would go on to become one of the most iconic automobiles in history. Designed by Malcolm Sayer, the E-Type was a masterpiece of automotive design, with its long, flowing lines and aggressive stance. But the E-Type wasn't just a pretty face; it was a true performance machine. Powered by a 3.8-liter engine, the E-Type could reach speeds in excess of 150 mph, making it one of the fastest cars of its time.

The E-Type captured the imagination of the public and car enthusiasts worldwide, becoming an instant symbol of the 1960s. Its beauty, combined with exceptional performance, helped to elevate Jaguar's image as a maker of world-class sports cars. It was also affordable compared to other supercars of the era, making it accessible to a wider range of buyers. The E-Type remains one of the most celebrated cars in automotive history, a true icon of design and performance.

The XJ Series: A Legacy of Luxury Sedans

While Jaguar's sports cars garnered much of the attention, the company also had a strong presence in the luxury saloon market. The Jaguar XJ series, which debuted in 1968, was one of the most important models in Jaguar's history. It represented the pinnacle of luxury and refinement, combining the best of Jaguar's performance heritage with cutting-edge technology and design.

The XJ was designed by Sir William Lyons, who wanted to create a car that was both beautiful and practical, with a focus on comfort and luxury. The car featured a refined, elegant design, a spacious and opulent interior, and powerful engines that made it a pleasure to drive. Over the years, the XJ series evolved, with each new generation building upon the legacy of its predecessors. The XJ became synonymous with luxury and class, attracting high-profile customers, including royalty, celebrities, and business leaders.

The XJ220: Jaguar's Supercar Dream

In the 1990s, Jaguar sought to break into the world of supercars with the XJ220, a car that would become the fastest production car in the world at the time. The XJ220 was a bold attempt by Jaguar to compete with the likes of Ferrari and Lamborghini. It was powered by a 3.5-liter twin-turbo V6 engine that could push the car to speeds of over 200 mph. With its futuristic design and blistering performance, the XJ220 was a true supercar, although its production was limited.

The XJ220's success was somewhat tempered by production issues and a limited number of units, but it remains an important part of Jaguar's history. It was a demonstration of Jaguar's ambition and its ability to push the boundaries of automotive engineering, even if the market for such a high-performance vehicle was more niche.

Chapter 4: Jaguar's Pioneering Role in British Automotive Design

Jaguar's reputation for combining style with performance has been a defining characteristic throughout its history, and much of that reputation stems from its pioneering role in automotive design. The brand's commitment to beautiful, functional design has made its cars some of the most visually striking vehicles on the road. But Jaguar's contribution to automotive design goes far beyond aesthetics; the brand has consistently pushed the boundaries of engineering and technology, creating cars that are as innovative as they are beautiful.

The Art of Automotive Design: Sir William Lyons' Vision

At the heart of Jaguar's design philosophy was its founder, Sir William Lyons. A visionary with a keen eye for style and a deep understanding of automotive engineering, Lyons set the tone for Jaguar's design language. From the beginning, Lyons was determined to create cars that were not just functional but works of art. His design philosophy was rooted in the belief that a car should be as visually appealing as it was practical.

Lyons' influence can be seen in every Jaguar model, from the early SS Jaguars to the iconic E-Type. His emphasis on sleek, aerodynamic lines and harmonious proportions made Jaguar's cars stand out from the competition. But Lyons was not just concerned with how a car looked; he also wanted his vehicles to offer a superior driving experience. His designs were always focused on combining beauty with performance, a combination that became the hallmark of the Jaguar brand.

Aerodynamics and Performance: The Jaguar Edge

One of the key elements of Jaguar's design philosophy was the focus on aerodynamics. From the very beginning, Jaguar engineers understood the importance of reducing drag and maximizing performance. This commitment to aerodynamics was evident in the design of models like the XK120, which featured a streamlined body that was not only beautiful but also functional. The car's low drag coefficient allowed it to reach high speeds, making it one of the fastest cars of its time.

Jaguar's focus on aerodynamics continued with the E-Type, which is often considered one of the most beautiful cars ever made. The E-Type's long, flowing lines and low, wide stance were designed not only for aesthetic appeal but also to enhance performance. The car's shape was carefully crafted to reduce drag and improve handling, allowing it to achieve speeds that were previously thought impossible for a car of its price range.

Jaguar's commitment to aerodynamics extended beyond individual models. The brand was also at the forefront of developing new manufacturing techniques and materials that would allow for lighter, more efficient vehicles. This focus on engineering innovation helped Jaguar to stay ahead of the competition and continue to produce cars that were both beautiful and high-performing.

Luxury and Comfort: The Jaguar Experience

In addition to performance and aerodynamics, Jaguar's design philosophy also placed a strong emphasis on luxury and comfort. From the beginning, Jaguar aimed to create cars that provided a superior driving experience, with interiors that were as carefully crafted as the exteriors. The use of high-quality materials, such as leather, wood, and polished metal, helped to create an atmosphere of sophistication and luxury in every Jaguar vehicle.

The introduction of the Jaguar XJ series in 1968 marked a new chapter in the brand's commitment to luxury and comfort. The XJ was designed to be a luxury saloon that offered all the performance of a sports car while providing the comfort and refinement expected of a high-end vehicle. The XJ set new standards for luxury saloons, with its spacious interior, high-tech features, and smooth, quiet ride. It quickly became the choice of royalty, celebrities, and business leaders, cementing Jaguar's reputation as a maker of world-class luxury cars.

Chapter 5: The Jaguar XJ Series: A Legacy of Luxury Sedans

The Jaguar XJ series, which debuted in 1968, represents one of the most important milestones in Jaguar's long history. Not only did it redefine what a luxury saloon could be, but it also solidified Jaguar's place as a leader in the luxury automotive market. The XJ was a car that combined performance, elegance, and technology into a package that was ahead of its time. Over the years, the XJ series has become a symbol of refinement and sophistication, and its legacy continues to this day.

The Birth of the XJ: A Vision Realized

The Jaguar XJ was the brainchild of Sir William Lyons, the company's founder, who was known for his keen sense of design and a deep understanding of automotive engineering. Lyons wanted to create a car that would be the pinnacle of luxury and performance, a car that would rival the best in the world while still embodying the distinctive elegance of Jaguar's design language. To achieve this, he tasked Jaguar's engineers with creating a luxury sedan that would not only offer exceptional comfort and refinement but also deliver the performance Jaguar was known for.

The first XJ model, the Series 1, was a breakthrough in design. The exterior featured smooth, flowing lines, and its sleek, modern look was a departure from the more traditional saloon designs of the time. The interior was equally impressive, with luxurious materials, spacious seating, and cutting-edge technology. Powered by Jaguar's XK engine, the XJ Series 1 offered a smooth and powerful ride that made it a pleasure to drive, while its advanced suspension system ensured a comfortable, quiet experience for passengers.

XJ Series Evolution: Innovation Meets Tradition

As the years passed, the XJ series continued to evolve, maintaining its reputation for luxury while embracing the latest technological advances. The Series 2, introduced in 1973, featured improvements in both performance and comfort, with a more refined interior and better handling. The Series 3, which debuted in 1979, took the luxury sedan to new heights, offering even more advanced features, such as electronic controls and a more powerful engine.

Throughout its history, the XJ series has consistently combined the latest in automotive technology with the brand's traditional focus on craftsmanship and luxury.

The series has featured numerous innovations, such as advanced suspension systems, improved engine technology, and high-end audio systems, while also maintaining the elegant design and superior driving experience that Jaguar is known for.

Despite its luxury status, the XJ series has always been a car that is fun to drive. Jaguar's engineers have continually focused on performance, ensuring that each iteration of the XJ not only delivers a smooth, comfortable ride but also provides exceptional handling and agility. The XJ's combination of power, comfort, and technology has made it the ideal choice for those seeking a luxury sedan that doesn't compromise on driving dynamics.

The XJ's Global Appeal: A Car for the Elite

The XJ's appeal has always extended far beyond the United Kingdom. Over the decades, the car has become synonymous with luxury and sophistication around the world. It has been favored by heads of state, business leaders, and celebrities, who appreciate its elegance, performance, and understated luxury. The XJ's presence in popular culture, from appearances in films and television to being the car of choice for prominent individuals, has helped cement its place as one of the most iconic luxury sedans in automotive history.

From the very first Series 1 to the latest generation of the XJ, the model has maintained its status as Jaguar's flagship sedan, representing the pinnacle of British automotive excellence. Its influence on the luxury sedan market is immeasurable, with competitors often trying to replicate its combination of style, performance, and luxury. The XJ series has been, and continues to be, the embodiment of everything Jaguar stands for: a perfect fusion of beauty, engineering, and refinement.

Chapter 6: The Birth of the E-Type: The Icon of the 1960s

When the Jaguar E-Type was unveiled in 1961, it didn't just change the automotive world; it redefined the concept of what a sports car could be. Often considered one of the most beautiful cars ever made, the E-Type was a perfect blend of form and function. It combined jaw-dropping design with exceptional performance, making it a true icon of the 1960s and solidifying Jaguar's position as a manufacturer of world-class sports cars.

The Vision Behind the E-Type

The idea for the E-Type came from Sir William Lyons, who wanted to create a car that could compete with the best in the world while also being affordable. At the time, Jaguar was already known for its high-performance cars, such as the XK120 and XK140, but Lyons was determined to take it to the next level. He envisioned a car that would not only offer exceptional speed and handling but also feature a design that would make it stand out in a crowded market of sports cars.

To bring this vision to life, Jaguar enlisted the help of renowned automotive designer Malcolm Sayer. Sayer, who had previously worked on the XK120 and other Jaguar models, was known for his work with aerodynamics and his ability to create cars that were both beautiful and functional. The result of this collaboration was the E-Type: a car that combined stunning good looks with blistering performance.

Design and Innovation: The E-Type's Impact

The E-Type was revolutionary in terms of both design and technology. Its sleek, aerodynamic shape was a radical departure from the angular lines of other sports cars of the era. The long, flowing curves of the body, the low, wide stance, and the aggressive front end all contributed to a design that was both beautiful and functional. The E-Type's design was so ahead of its time that Enzo Ferrari, the founder of Ferrari, famously called it "the most beautiful car ever made."

Under the hood, the E-Type was just as impressive. The car was powered by a 3.8-liter inline-six engine, capable of producing 265 horsepower. This gave the E-Type a top speed of 150 mph, making it one of the fastest cars of its time. The car's lightweight construction and advanced suspension system ensured that it handled exceptionally well, providing a driving experience that was both exhilarating and refined.

The E-Type was also one of the most affordable sports cars of its era. While cars from Ferrari and Porsche were priced out of reach for most buyers, the E-Type offered similar performance and design at a fraction of the cost. This made it accessible to a wider audience and helped cement Jaguar's reputation as a maker of performance cars that were not only beautiful but also practical.

Cultural Icon: The E-Type and the 1960s

The E-Type quickly became a symbol of the 1960s, embodying the spirit of the era with its cutting-edge design and performance. It was featured in numerous films and television shows, becoming a symbol of cool sophistication and youthful rebellion. Its sleek lines and powerful performance made it a favorite of celebrities and car enthusiasts alike, and it was soon regarded as one of the most iconic cars in automotive history.

The E-Type's popularity was not just confined to its appearance on the silver screen. It also became a symbol of freedom and independence, representing the aspirations of a generation that was coming of age in a time of social and cultural change. The E-Type was a car that captured the spirit of the 1960s, and its enduring appeal continues to this day.

Jaguar produced several versions of the E-Type over the years, including the Series 2 and Series 3, each offering improvements in performance and comfort. However, it was the original Series 1, with its stunning design and blistering performance, that became the most iconic and remains a benchmark for Jaguar's legacy.

Enduring Legacy: The E-Type's Place in History

The E-Type's impact on the automotive world cannot be overstated. It not only helped cement Jaguar's reputation as a maker of world-class sports cars, but it also set a new standard for design and performance in the automotive industry. The E-Type's legacy continues to influence car design today, and it remains one of the most celebrated cars in automotive history. Its timeless beauty, combined with its groundbreaking performance, ensures that the E-Type will always be remembered as one of the greatest cars ever made.

Chapter 7: Jaguar's Racing Heritage: From Le Mans to Formula 1

Jaguar's reputation for performance is not confined to the world of road cars. The company's motorsport heritage is deeply woven into the fabric of its history, with the brand consistently achieving remarkable success on racetracks around the world. From the demanding endurance races at Le Mans to Formula 1's cutting-edge technological battles, Jaguar has made a lasting impact on motorsports, using its track experience to refine its cars and push the limits of performance and innovation.

Early Racing Success: Le Mans and Beyond

Jaguar's motorsport journey began in the 1950s, just as the brand was solidifying its position in the luxury car market. Early success came in the form of the XK120 and later the XK140, which were used in various racing events, notably the prestigious 24 Hours of Le Mans. However, it was the legendary D-Type, introduced in 1954, that would secure Jaguar's place in racing history.

The D-Type was designed specifically for Le Mans, a grueling endurance race that demanded a perfect blend of speed, endurance, and handling. With its sleek aerodynamic design and a powerful 3.4-liter straight-six engine, the D-Type was built to dominate. It proved to be an unstoppable force, winning the 24 Hours of Le Mans three times in a row from 1955 to 1957, cementing Jaguar's dominance in endurance racing.

Jaguar's success at Le Mans was not just a matter of having the best car on the track; it was also a testament to the company's innovative engineering. The D-Type was one of the first cars to incorporate a monocoque construction, a lightweight, strong design that was crucial to achieving the speeds required at such a demanding race. This engineering breakthrough would have a lasting impact on automotive design and set the stage for many of Jaguar's future innovations.

The E-Type and the Racing Legacy

While the E-Type is best known for its contributions to road cars, it also had a significant presence in racing. The E-Type's combination of performance, handling, and sleek design made it a natural fit for motorsport. Jaguar's racing success with the E-Type came primarily in the form of privateer efforts, with many owners campaigning the car in various racing series, from

SCCA (Sports Car Club of America) events to European GT races.

Though not factory-supported like the D-Type, the E-Type's involvement in motorsports helped solidify its reputation as a high-performance machine. Jaguar's motorsport success in the 1960s with the E-Type continued to reinforce its image as a brand that could deliver cutting-edge performance on both the road and the track.

Formula 1 and the Modern Era

Jaguar's commitment to motorsport extended well into the 21st century. In 2000, the brand entered Formula 1 with the launch of the Jaguar Racing team, a move that sought to position the brand at the forefront of global motorsport competition. The team, which was initially formed from the remnants of the Stewart Grand Prix team, entered the high-stakes world of Formula 1 with lofty aspirations.

Though Jaguar Racing did not achieve the immediate success it hoped for, its involvement in F1 showcased the brand's technical prowess and its desire to compete at the highest levels. Jaguar's F1 team worked with advanced engineering and developed innovative technologies that helped further the brand's reputation for pushing the boundaries of performance. The team's involvement in F1 was relatively short-lived, with Jaguar eventually selling its team to Red Bull Racing in 2004, but the experience helped Jaguar further refine its engineering and performance capabilities.

In addition to Formula 1, Jaguar's racing heritage has also extended into other motorsport disciplines, including touring car racing and endurance racing. The company's focus on motorsport has not only helped it develop high-performance vehicles but also allowed it to demonstrate its commitment to excellence in engineering.

A Legacy of Innovation

Jaguar's involvement in motorsport has had a profound effect on its road cars. The technologies, engineering practices, and design philosophies developed on the track have often been transferred to Jaguar's production cars, resulting in innovations such as advanced suspension systems, lightweight materials, and aerodynamic enhancements. The D-Type's monocoque construction, for example, would influence the development of future Jaguar cars, including the iconic E-Type.

The company's racing heritage remains a key part of its identity. Whether it's the Le Mans victories of the D-Type, the global success of the E-Type in motorsports, or the technical achievements in Formula 1, Jaguar's motorsport legacy is one of innovation, performance, and a relentless pursuit of excellence. This heritage continues to shape the company's engineering and design philosophy, ensuring that Jaguar remains a performance brand at the heart of the automotive world.

Chapter 8: Jaguar and the British Motor Industry: A Pillar of Excellence

Jaguar's history is inseparable from the broader story of the British motor industry. Since its inception, the brand has played a key role in shaping not only the nation's automotive landscape but also its cultural and economic identity. As one of Britain's most storied and prestigious car manufacturers, Jaguar has become a symbol of the country's engineering ingenuity, design excellence, and craftsmanship.

A British Icon: The Heart of British Automotive Heritage

From its early days as the Swallow Sidecar Company, Jaguar has been deeply intertwined with the British motor industry. The company's success and contributions to the automotive world helped establish Britain as a global center for car manufacturing. In the 1950s and 1960s, Jaguar was one of the leaders in the British automotive renaissance, producing cars that were not only technologically advanced but also aesthetically groundbreaking.

Jaguar's reputation for creating luxurious, high-performance cars positioned it as a flagship of British automotive engineering. Its early success, particularly in motorsport with the D-Type, demonstrated that British carmakers could compete with the best in the world, while the E-Type became a cultural symbol of the 1960s. As Jaguar continued to innovate and refine its products, it embodied the values of British craftsmanship—elegant design, superb engineering, and a deep-rooted sense of tradition.

The British Leyland Era: Challenges and Change

Despite its success, Jaguar's journey through the British automotive industry was not without its challenges. In the late 1960s, Jaguar became part of the British Leyland Motor Corporation, a move that marked a significant turning point in the company's history. British Leyland, which was formed by the merger of several British car manufacturers, struggled with labor issues, financial instability, and declining quality control. This period saw Jaguar's reputation suffer, and the company faced difficulties in maintaining its position as a luxury brand.

During this time, Jaguar's engineering prowess continued to shine through in its products, with models like the XJ Series and the Jaguar XJ-S (introduced in the

1970s), but the internal struggles of British Leyland placed a strain on the company's operations. The decline of the British Leyland group in the 1980s led to the eventual privatization of Jaguar, allowing the company to regain its independence and once again focus on its legacy as a maker of luxury, high-performance cars.

A Revived Jaguar: A Return to Prestige

In 1984, Jaguar was sold to the Ford Motor Company, which gave the brand a new lease on life. Under Ford's ownership, Jaguar was able to invest in new technologies, improve its manufacturing processes, and rebuild its reputation as a producer of premium vehicles. The 1990s saw the introduction of models such as the XK8, a modern take on Jaguar's sports car legacy, and the S-Type, which combined luxury with the latest in automotive technology.

Jaguar's rebirth continued in the 21st century, especially following its acquisition by Tata Motors in 2008. Under Tata, Jaguar regained its status as a symbol of British luxury and innovation. The brand launched a range of new vehicles, including the XF, XJ, and F-Type, and began focusing heavily on design, performance, and sustainability. Jaguar's commitment to producing world-class vehicles ensured that it remained an important player in the global automotive industry.

Jaguar and the Future of British Motoring

Today, Jaguar stands as a key pillar of the British motor industry, representing the best of British design and engineering. While the landscape of the industry has changed with the rise of new technology and shifting global markets, Jaguar continues to thrive by embracing innovation, sustainability, and a renewed focus on performance.

As the automotive world shifts towards electric vehicles, Jaguar has embraced the challenge of staying at the forefront of the industry. With a renewed focus on electric mobility, Jaguar is positioning itself as a leader in luxury electric cars, ensuring that its rich legacy of performance and design continues into the future. The brand's new range of electric vehicles, such as the I-PACE, represents a bold new direction while retaining the core values that have made Jaguar an iconic British brand.

Jaguar's role as a pillar of the British motor industry is

secure, and its legacy continues to shape the future of motoring in the UK and around the world. As the company moves forward, it remains committed to excellence in design, performance, and innovation, ensuring that it will remain a key player in the ever-evolving world of automobiles.

Chapter 9: The Changing Landscape: Mergers, Acquisitions, and Ownership

Jaguar's history has been marked by a series of shifts in ownership and strategic decisions that have shaped the company's direction and future. The fluctuating nature of ownership in the automotive industry, combined with the ever-changing demands of global markets, meant that Jaguar had to navigate through times of uncertainty, adaptation, and transformation. From its early days as an independent company to its various mergers and acquisitions, the story of Jaguar's ownership is one of resilience, reinvention, and survival.

British Leyland Era: Merging for Survival

In the 1960s, Jaguar became part of the British Leyland Motor Corporation, a move that was intended to safeguard the company's future in a rapidly consolidating British automotive industry. British Leyland, a conglomerate that merged several major car brands, including Austin, Morris, and MG, was created in 1968 in response to increasing global competition. Jaguar's inclusion in British Leyland marked a dramatic shift for the company.

While British Leyland provided Jaguar with the financial backing and scale to survive in a challenging environment, the integration of so many different brands proved to be problematic. The conglomerate struggled with management issues, labor disputes, and quality control problems, all of which affected Jaguar's production and reputation. The lack of focused leadership and the pressure of operating within a larger, disorganized group led to Jaguar's struggles in the 1970s and early 1980s.

Despite these challenges, Jaguar continued to produce iconic cars, such as the XJ Series and the XJ-S, during this period. However, British Leyland's lack of cohesion prevented Jaguar from truly reaching its potential. The 1980s would be a time of transition for Jaguar, as the company sought to regain its independence and status as a world-class manufacturer of luxury vehicles.

Ford Acquisition: A New Chapter for Jaguar

In 1984, Jaguar was privatised and acquired by the Ford Motor Company. This acquisition marked a turning point for the brand, as Ford brought the stability, resources, and global reach that Jaguar needed to recover from its previous struggles. Under Ford's ownership, Jaguar was able to focus on developing new vehicles, improving its manufacturing processes, and regaining its reputation for quality and performance.

During the 1990s, Ford made significant investments in Jaguar, resulting in the introduction of new models such as the XK8, a modern version of Jaguar's sports car lineage, and the S-Type, which successfully combined luxury with technology. The Ford era also saw Jaguar expand its presence in the global market, particularly in the United States, where it became a major player in the luxury sedan market.

One of the most significant outcomes of Ford's ownership was Jaguar's ability to continue its tradition of producing performance-oriented vehicles while integrating modern technology and manufacturing techniques. Ford's influence ensured that Jaguar could offer cars that were both competitive and technologically advanced, without compromising on the craftsmanship and luxury that the brand had become known for.

However, despite the successes under Ford, the brand was still searching for its own identity in a rapidly changing automotive world. Ford's own challenges, including the global financial crisis in 2008, led to a shift in strategy, and Jaguar's future would take another dramatic turn with its acquisition by Tata Motors.

Tata Motors: A Global Strategy for the Future

In 2008, Jaguar was sold to Tata Motors, an Indian automotive giant, in a deal that was initially met with mixed reactions. For many, this was seen as the end of an era for Jaguar as a quintessentially British brand. However, Tata Motors' ownership would breathe new life into the brand and help it navigate the challenges of the 21st century.

Under Tata's ownership, Jaguar was able to re-establish itself as a premium, luxury brand while also embracing new technology and innovation. Tata's financial backing allowed Jaguar to invest in new models and enter emerging markets, particularly in Asia, where the brand began to grow its presence.

Jaguar's first major success under Tata was the launch of the XF sedan in 2008. The XF was a revolutionary model that combined modern design with Jaguar's signature luxury and performance. The car received widespread acclaim for its stylish exterior, high-tech interior, and dynamic driving experience, reinvigorating the brand's image and signaling a new direction for Jaguar.

Tata also made a significant commitment to sustainability, leading Jaguar to focus on environmentally-friendly technology. This shift culminated in the development of the all-electric I-PACE SUV in 2018, Jaguar's first fully electric vehicle. The I-PACE proved to be a major success, winning numerous awards and establishing Jaguar as a leader in the electric vehicle (EV) space.

Jaguar's transformation under Tata Motors proved that it was possible for the brand to stay true to its heritage while embracing the future of mobility. Tata's ownership has been instrumental in positioning Jaguar as a forward-thinking, global player in the luxury automotive market.

Chapter 10: Design and Innovation: The Jaguar Approach to Crafting Cars

Jaguar has long been celebrated not just for its performance but also for its distinctive design and innovative engineering. The company's approach to automotive design is rooted in a philosophy that combines style, functionality, and cutting-edge technology. From the sleek lines of the E-Type to the modern, bold look of the F-Type, Jaguar's design language has consistently been about creating cars that are as visually striking as they are technologically advanced. At the core of Jaguar's success is the belief that a car should be more than just a means of transportation—it should be a work of art.

Form Meets Function: The Role of Design in Jaguar's Legacy

One of the key elements of Jaguar's design philosophy is the belief that form and function must work in harmony. From the early days of the company, Sir William Lyons emphasized the importance of creating cars that were not only beautiful but also practical and performant. The result has been a series of cars that not only look stunning but also offer superior handling, comfort, and driving pleasure.

The design of the E-Type, for example, is often regarded as one of the most beautiful car designs ever created. Its long, sweeping curves were not just aesthetically pleasing; they were also functional, enhancing the car's aerodynamics and performance. The E-Type's design combined the best of form and function, making it both a technical marvel and a visual icon. This philosophy of designing cars that were both beautiful and efficient would become a hallmark of Jaguar's approach to automotive engineering.

Jaguar's commitment to combining design with performance can be seen in all of its vehicles, from the luxurious XJ sedans to the modern F-Type sports car. Whether it's the use of lightweight materials, advanced suspension systems, or innovative aerodynamics, Jaguar's design is always focused on creating cars that deliver a thrilling driving experience while also being visually captivating.

Innovative Technology: Pushing the Boundaries of Performance

Jaguar has consistently pushed the boundaries of

automotive engineering, introducing groundbreaking technologies that have set new standards in the industry. In the 1950s, the D-Type's monocoque construction revolutionized the way cars were built, allowing for a lighter, stronger, and more aerodynamically efficient design. This innovation not only helped Jaguar win three consecutive Le Mans races but also influenced the design of future road cars.

In the 21st century, Jaguar continued to innovate with the introduction of cutting-edge technologies such as the lightweight aluminum architecture used in the F-Type, the advanced suspension systems in the XJ, and the integration of hybrid and electric powertrains in its vehicles. Jaguar's focus on innovation has helped the brand remain at the forefront of the luxury automotive market, ensuring that its cars are always equipped with the latest in performance, comfort, and sustainability.

The Future of Jaguar Design: Electrification and Sustainability

As the automotive industry shifts toward electrification, Jaguar has embraced this change while staying true to its design principles. The launch of the I-PACE, Jaguar's first all-electric vehicle, marked a new era for the brand, showcasing its commitment to sustainability without sacrificing the performance and luxury for which Jaguar is known. The I-PACE's futuristic design, advanced electric powertrain, and exceptional driving dynamics make it a true Jaguar, proving that the brand can lead the way in the electric age while maintaining its heritage of design excellence.

Looking ahead, Jaguar is continuing to evolve its design language, incorporating more sustainable materials, innovative technologies, and a focus on reducing its environmental impact. The company's commitment to creating vehicles that are both beautiful and sustainable ensures that Jaguar will remain a key player in the future of the automotive industry.

Chapter 11: Jaguar and the American Market: A Tale of Love and Challenges

Jaguar's relationship with the American market has been a defining part of the brand's history, characterized by both triumphs and struggles. In the post-war era, as the American automotive market began to shift toward more luxurious and performance-oriented vehicles, Jaguar seized the opportunity to expand its presence in the United States. However, despite the allure of British luxury and the desire for sophisticated performance cars, Jaguar's journey in the U.S. has been anything but straightforward.

The Early Years: A Growing Reputation

Jaguar's introduction to the American market in the 1950s was a pivotal moment in the brand's history. At the time, American consumers were largely dominated by domestic brands, with Detroit's Big Three automakers—Ford, General Motors, and Chrysler—leading the market. European cars, particularly from Germany and Italy, were seen as luxury items and exotic imports. Jaguar, with its British heritage and focus on high-end performance, found a ready audience among wealthy American buyers looking for something different, something more refined.

The XK120, introduced in 1948, was one of the first Jaguars to make a significant impact in the United States. The car's stunning design, combined with its performance credentials, made it an instant hit. The XK120 was a car that could rival the best American-made luxury vehicles while offering something that American automakers couldn't: European craftsmanship and performance. As Jaguar continued to expand its lineup, it found success with models like the XK140 and XK150, further solidifying its reputation in America.

The real breakthrough for Jaguar in the U.S. came with the launch of the E-Type in 1961. The E-Type was nothing short of revolutionary in terms of design, performance, and affordability. With its sleek, aerodynamic lines and high-speed capabilities, the E-Type captured the American imagination and became a symbol of youthful rebellion and sophistication. The E-Type's combination of beauty and performance was perfectly aligned with American tastes in the 1960s, and the car quickly became one of the most desirable vehicles on the market. Its presence in pop culture, especially in Hollywood, helped cement Jaguar's status as the brand to own for the American elite.

The Challenges of Expansion: Reliability and Service Issues

While Jaguar found success in the United States, the brand's relationship with the American market was not without its challenges. As Jaguar expanded its operations, it faced growing pains in terms of quality control, production capacity, and customer service. Many American buyers were drawn to the Jaguar experience, but they were often frustrated with the brand's reliability issues.

Jaguar's cars, although beautifully designed, were notorious for having mechanical problems. Electrical issues, rust problems, and cooling system failures were not uncommon in early Jaguar models. This led to a growing perception in the U.S. that while Jaguar produced stunning vehicles, they were plagued by reliability issues. The brand's reliance on hand-built craftsmanship, while adding to the exclusivity of the cars, also meant that manufacturing processes were more prone to inconsistency.

In addition to mechanical challenges, Jaguar faced difficulties with its dealership network in the U.S. Early dealers struggled to provide the level of customer service that American buyers expected from a luxury brand. The lack of readily available parts, long wait times for repairs, and less-than-ideal customer service contributed to the growing frustration among Jaguar owners.

Despite these challenges, Jaguar's allure never completely waned. The brand continued to be associated with British elegance and performance, and there remained a loyal base of customers who appreciated the brand's distinctiveness. Jaguar's designs, such as the XJ and the XJS, remained iconic, and the appeal of owning a Jaguar persisted throughout the decades.

The 1990s: A New Approach and Renewed Success

By the 1990s, Jaguar was well-established in the U.S. market, but it was clear that the brand needed to evolve to regain its position as a leader in the luxury market. Ford's acquisition of Jaguar in 1989 played a pivotal role in the company's ability to address some of the issues it faced in the U.S. market. Ford's financial backing allowed Jaguar to modernize its manufacturing process, improve quality control, and address reliability concerns.

The introduction of the Jaguar XJ6 in the 1990s was a significant step toward revitalizing the brand's image in America. The XJ6 was a more refined, reliable, and comfortable car that retained the design elegance Jaguar was known for. The 1990s also saw the launch of the XK8, which reinvigorated the brand's sports car heritage and helped Jaguar maintain a strong presence in the American market.

Jaguar's efforts to improve its reputation for reliability paid off in the 2000s. The brand continued to expand its lineup, introducing models like the S-Type and the XF, both of which were well-received by American buyers. By the 2010s, Jaguar was regarded as a solid contender in the luxury market, offering cars that combined stunning design, advanced technology, and dependable performance.

A Future of Electrification and Innovation

In recent years, Jaguar's strategy has shifted toward embracing new technologies, including electrification, to remain relevant in the highly competitive American market. The introduction of the I-PACE, Jaguar's first all-electric vehicle, marked a significant step in the company's future plans. The I-PACE, with its blend of performance, sustainability, and luxury, has received critical acclaim and is a testament to Jaguar's ability to adapt to changing consumer preferences.

As Jaguar continues to evolve in the U.S. market, it is committed to offering vehicles that meet the demands of modern buyers, particularly those seeking eco-friendly alternatives without sacrificing luxury or performance. The future of Jaguar in America is closely tied to its continued investment in electric vehicles and its ability to remain at the forefront of design and innovation.

Chapter 12: The XJ220: The Supercar That Almost Was

The Jaguar XJ220 remains one of the most intriguing and enigmatic chapters in the company's history. Produced in the 1990s, the XJ220 was Jaguar's bold foray into the world of hypercars, a category dominated by manufacturers like Ferrari and Lamborghini. The XJ220 was intended to be a game-changer, a car that would not only showcase Jaguar's engineering prowess but also establish it as a serious contender in the supercar market. While the car would go on to become one of the fastest production cars in the world at the time, its troubled production process and limited success on the market make it a symbol of ambition, frustration, and missed opportunities.

The Vision for the XJ220

The idea for the XJ220 was conceived in the late 1980s when Jaguar engineers, under the leadership of Ford, began to explore the idea of creating a supercar that would rival the best in the world. At the time, the automotive world was obsessed with performance and speed, and Jaguar saw an opportunity to create a car that would embody the brand's performance heritage while pushing the boundaries of automotive engineering.

Jaguar first unveiled the XJ220 as a concept car at the 1988 Birmingham Motor Show. The car was a revelation, with its futuristic design and remarkable performance figures. The concept car was powered by a V12 engine and promised to deliver speeds exceeding 220 mph, a truly remarkable feat for a car of its time. The public was captivated by the XJ220, and Jaguar received an overwhelming amount of interest from potential buyers, with over 1,000 expressions of interest.

The Production Process and Controversy

Despite the initial excitement surrounding the XJ220, the road to production was fraught with issues. Jaguar, in its pursuit of creating the ultimate supercar, faced numerous technical challenges. The V12 engine that had been showcased in the concept car proved to be too difficult and expensive to mass-produce. In order to meet production deadlines and reduce costs, Jaguar decided to replace the V12 with a twin-turbocharged V6 engine, a decision that would later prove to be controversial.

The V6 engine, while powerful and capable of

producing significant horsepower, was not what buyers had expected from a Jaguar supercar. The decision to change the engine upset many potential customers, who felt that Jaguar had compromised on the car's initial promise. The XJ220's production delays, coupled with the change in engine specification, led to frustration among the early buyers, many of whom had already put down significant deposits.

Despite the controversies surrounding the XJ220, the car that eventually emerged was an extraordinary machine. Powered by a 3.5-liter twin-turbo V6 engine, the XJ220 was capable of reaching a top speed of 217 mph, making it one of the fastest production cars in the world at the time. It was equipped with advanced technologies, including an all-wheel-drive system and carbon fiber components, which made it a true marvel of engineering.

Legacy of the XJ220

Unfortunately, the XJ220's troubled production and controversial specifications led to its underwhelming success in the market. Jaguar produced only 271 units of the XJ220, far fewer than the 1,500 units that had been initially projected. The car's exclusivity, coupled with its high price tag and the ongoing controversy, limited its appeal to a select group of wealthy buyers.

Today, the XJ220 is regarded as one of Jaguar's greatest achievements, despite its commercial struggles. It remains a sought-after collector's item, admired for its cutting-edge technology and performance. The car's legacy is a reminder of Jaguar's ambition to compete with the world's best in the supercar market, even though it didn't achieve the commercial success the company had hoped for.

The XJ220 serves as a testament to Jaguar's engineering excellence and its willingness to push the limits of what was possible in automotive design. While the XJ220 may not have been the massive success that Jaguar envisioned, it remains a significant part of the company's legacy and a symbol of the brand's pursuit of performance and innovation.

Chapter 13: The Revival of Jaguar: The 21st Century and New Challenges

Jaguar's entry into the 21st century marked a period of transition, as the brand sought to adapt to the rapidly evolving automotive landscape. After a turbulent history of mergers, acquisitions, and changes in ownership, Jaguar was entering an era where global competition, environmental concerns, and shifting consumer tastes were reshaping the way cars were made and sold. Yet, despite these challenges, Jaguar's heritage of luxury, performance, and elegance remained firmly in place, driving the brand's revival.

Ford's Influence and the New Millennium

In 1989, Ford purchased Jaguar, and under its ownership, the brand received the stability and financial backing it needed to modernize. By the early 2000s, Ford had invested heavily in Jaguar's product line and manufacturing, leading to the introduction of several key models, including the stylish S-Type sedan and the highly anticipated XK8 sports car. These vehicles marked a return to the more traditional elements of Jaguar's identity, with a renewed focus on luxury, performance, and refinement.

However, the turn of the century also brought significant challenges. The global automotive industry was undergoing rapid transformation, with new technologies and changing customer preferences shaping the market. Jaguar needed to ensure it could compete not only with established European rivals like Mercedes-Benz and BMW but also with a rising wave of luxury performance brands from around the world.

The year 2000 saw the release of the XK8, a modern reinterpretation of Jaguar's classic sports car design. The XK8 was well-received, offering a combination of advanced engineering, sharp handling, and sleek design. The introduction of the XJ6 sedan further bolstered Jaguar's position as a luxury automaker with a strong balance of performance and comfort. However, it was clear that Jaguar had to overcome certain issues, particularly regarding product quality, to maintain its reputation as a world-class brand.

The Tata Motors Era: A New Beginning

In 2008, Ford sold Jaguar, along with Land Rover, to Tata Motors, an Indian automotive giant with the resources and vision to lead Jaguar into a new chapter. At first, many questioned whether Tata would be able to maintain Jaguar's British heritage while taking it into a more modern, competitive age. However, Tata's ownership would prove to be a turning point for the company.

Under Tata Motors, Jaguar experienced a resurgence in both design and innovation. Tata's financial backing allowed Jaguar to invest in new technologies, refine its products, and expand its global presence. One of the most significant developments under Tata's ownership was the re-engineering of Jaguar's product portfolio to incorporate cutting-edge technology, sustainability, and contemporary design, ensuring that the brand remained relevant to a new generation of luxury buyers.

In 2008, Jaguar introduced the XF sedan, which became one of the brand's most successful and critical models. The XF's sleek, modern design, combined with the latest in automotive technology, helped the brand regain its standing in the luxury market. It was praised for its refined handling, comfort, and performance, making it a favorite among enthusiasts and consumers alike.

Jaguar also launched the F-Type sports car in 2013, a sleek, modern take on the company's iconic roadster lineage. The F-Type proved to be a true revival of

Jaguar's sports car credentials, offering exhilarating performance, striking good looks, and a driving experience that captured the essence of the brand's legacy. The F-Type, powered by a range of V6 and V8 engines, showcased Jaguar's ability to blend traditional performance with modern engineering.

The brand also began to focus on hybrid and electric technologies, particularly with the introduction of the I-PACE, Jaguar's first fully electric vehicle, in 2018. The I-PACE marked Jaguar's commitment to the future of sustainable mobility while remaining true to its performance roots. The all-electric I-PACE quickly gained attention for its impressive range, cutting-edge features, and thrilling driving experience, making it one of the leading electric vehicles in the luxury market.

Adapting to a Changing Automotive World

Despite the many successes under Tata's ownership, Jaguar faced new challenges in the 2020s. The shift towards electric vehicles, stricter environmental regulations, and the impact of the COVID-19 pandemic on global manufacturing presented hurdles for Jaguar. However, the brand's legacy of innovation, design excellence, and performance positioned it well to meet these challenges head-on.

Jaguar's move towards electrification is seen in its ambitious plans to become a fully electric brand by 2025, as part of a broader strategy to reduce its carbon footprint and lead the luxury car market into the future. By embracing cutting-edge technologies, Jaguar is ensuring that its brand values of luxury, performance, and style are preserved, even as the world shifts toward sustainable mobility.

As the brand enters its second century of existence, Jaguar is committed to reimagining the luxury car experience for a new generation of drivers. Through continuous innovation, design excellence, and a renewed focus on sustainability, Jaguar is poised to continue its legacy of creating world-class vehicles for the discerning driver.

Chapter 14: Jaguar's Contribution to Electric Vehicles: A New Era of Mobility

Jaguar's commitment to innovation has always been a central part of its brand identity, and in the 21st century, the company has embraced a new frontier in automotive technology: electric vehicles (EVs). As the automotive industry increasingly shifts toward sustainability, Jaguar has sought to maintain its position as a leader in luxury and performance while addressing the growing demand for cleaner, more efficient vehicles. With the launch of the I-PACE, Jaguar has firmly established itself as a key player in the electric mobility revolution.

A Bold Move Toward Electrification

In the mid-2010s, Jaguar recognized the need to diversify its product range and prepare for the future of mobility. With the rise of Tesla and the increasing regulatory pressures on automakers to reduce emissions, the time was ripe for Jaguar to make a bold move toward electric mobility. In 2018, the brand introduced the I-PACE, its first fully electric vehicle, marking a significant milestone in the company's long history.

The I-PACE was not just another electric car—it was a game-changer. Combining Jaguar's legacy of luxury and performance with cutting-edge electric technology, the I-PACE quickly garnered attention for its design, performance, and environmental benefits. Powered by a 90-kWh battery, the I-PACE offered a range of up to 234 miles on a single charge, along with rapid acceleration and a top speed of 124 mph. The I-PACE proved that electric vehicles could offer both environmental benefits and thrilling driving experiences, staying true to Jaguar's commitment to performance.

The I-PACE was also praised for its interior, which reflected Jaguar's focus on luxury and comfort. The cabin featured high-quality materials, modern infotainment systems, and an intuitive layout that made it clear that this was no compromise on the traditional Jaguar experience. The I-PACE's combination of performance, luxury, and cutting-edge technology helped elevate Jaguar's status as an electric vehicle leader in the luxury segment.

Sustainability and the Future of Electric Mobility

Jaguar's commitment to sustainability goes beyond the

production of the I-PACE. As part of the company's broader strategy, Jaguar has made significant strides in reducing its carbon footprint and minimizing the environmental impact of its operations. From its manufacturing processes to its supply chain management, Jaguar has sought to implement sustainable practices across its business.

The shift to electrification is part of a broader trend in the automotive industry, where traditional internal combustion engines are being replaced by electric powertrains in order to meet stricter environmental regulations and reduce dependence on fossil fuels. Jaguar has embraced this shift wholeheartedly, committing to becoming a fully electric car manufacturer by 2025. This transition is part of the brand's goal to meet global sustainability targets, reduce emissions, and offer customers more eco-friendly choices without sacrificing the performance and luxury that Jaguar is known for.

The move toward a fully electric future also aligns with Jaguar's commitment to cutting-edge technology and innovation. The company is investing heavily in electric vehicle research and development, working on new battery technologies, charging infrastructure, and other advancements that will help ensure the long-term success of its electric vehicles.

The Road Ahead: Leading the Charge in Electric Luxury

Jaguar's electric future represents a new era of mobility, one where performance, luxury, and sustainability go hand in hand. As the brand continues to evolve, it will undoubtedly face new challenges and competition in the growing electric vehicle market. However, Jaguar's rich heritage, commitment to innovation, and unwavering focus on luxury and performance give it a strong foundation for success.

The next chapter in Jaguar's journey will be defined by electric vehicles that continue to push the boundaries of what is possible in the automotive world. With the I-PACE as its flagship, Jaguar is ready to lead the charge in the electric revolution, ensuring that its legacy of creating world-class vehicles will endure for generations to come.

Chapter 15: The Role of Jaguar in Luxury and Performance Car Culture

Jaguar's identity has always been intertwined with the culture of luxury and performance, representing the epitome of British craftsmanship and engineering excellence. Since its inception, Jaguar has been more than just a car manufacturer—it has been a symbol of sophistication, taste, and driving passion. Its cars are not only designed to perform at the highest levels but also to provide an experience that goes beyond mere transportation. In this chapter, we explore how Jaguar has shaped and continues to influence the culture of luxury and performance cars.

Jaguar's Early Influence: The Dawn of Luxury Performance

From its earliest days, Jaguar established itself as a maker of cars that combined beauty, performance, and luxury. In the 1950s, the XK120, followed by the XK140 and XK150, set the standard for what a sports car could be. These models were sleek, powerful, and affordable compared to their European rivals, and they made Jaguar synonymous with speed, agility, and style. The XK120 was a revolution in terms of its design and speed, capturing the attention of both car enthusiasts and the general public alike.

The 1960s marked Jaguar's most significant leap into the luxury performance car market with the launch of the E-Type. Not only was the E-Type faster and more powerful than many of its competitors, but it also brought an entirely new aesthetic to the automotive world. Its streamlined, seductive design and elegant proportions were unlike anything the public had seen, establishing Jaguar as a true trendsetter. The E-Type was the car that epitomized Jaguar's unique blend of engineering brilliance and British luxury.

Jaguar's Role in Luxury Car Culture

Jaguar's role in luxury car culture cannot be overstated. From the elegance of the XJ sedan to the groundbreaking designs of the F-Type, Jaguar has continuously created vehicles that appeal to the discerning buyer who values both comfort and performance. The XJ, in particular, with its combination of superb handling, refined design, and technological sophistication, became the embodiment of the luxury sedan. It became a favorite of business leaders, celebrities, and royalty, providing a quiet, comfortable, yet exhilarating driving experience.

The key to Jaguar's success in the luxury car market has been its focus on quality and craftsmanship. Unlike mass-market manufacturers, Jaguar's cars have always been designed for those who seek more than just a mode of transportation. Whether it's the use of premium materials in the cabin or the finely tuned suspension systems that provide unparalleled comfort and handling, every aspect of a Jaguar is crafted with precision and care.

In the 21st century, Jaguar has continued to lead in the luxury car market with models like the XF and the F-Type. The XF sedan redefined what a mid-luxury car could be, blending advanced technology with a design that is both modern and timeless. The F-Type, meanwhile, has become a modern icon in the sports car world, offering a driving experience that is both thrilling and sophisticated. These models have helped Jaguar maintain its reputation as a leader in the luxury and performance sectors.

Jaguar's Performance DNA: From Le Mans to Road Cars

While Jaguar has always been associated with luxury, the brand's roots in motorsport have played a crucial role in shaping its performance ethos. The D-Type's triumphs at Le Mans in the 1950s demonstrated that Jaguar was more than just a maker of beautiful cars—it was a company that could create performance machines capable of competing with the best in the world. The engineering innovations developed for racing, such as the monocoque construction of the D-Type, would go on to influence Jaguar's production cars for decades.

The E-Type's performance credentials were also shaped by the company's racing heritage. Its powerful engine and lightweight design were a direct result of Jaguar's commitment to producing high-performance cars, and the E-Type was widely regarded as one of the fastest and most beautiful cars of its time. Today, Jaguar's performance models, such as the F-Type, continue to showcase the brand's racing legacy, with advanced technology, powerful engines, and precision handling ensuring that every drive is as thrilling as it is refined.

Jaguar's performance cars, from the XK120 to the F-Type, represent more than just speed—they symbolize the thrill of driving. Jaguar's ability to blend performance with luxury, comfort, and style has cemented its place as one of the most influential brands in the luxury performance car culture. The driving experience in a Jaguar is unlike any other—smooth, exhilarating, and always with an underlying sense of elegance and refinement.

Jaguar in the Age of Electric Vehicles: A New Era of Performance

As the automotive industry shifts towards electric vehicles, Jaguar has embraced the opportunity to redefine performance once again. The introduction of the I-PACE in 2018 marked a significant step in Jaguar's journey into the electric vehicle (EV) world, blending the brand's legacy of performance with the latest in electric powertrain technology. The I-PACE has been widely praised for its driving dynamics, proving that an electric car can be just as thrilling and engaging as a traditional petrol-powered vehicle.

Jaguar's commitment to electric performance is a key part of its strategy for the future. The brand's focus on creating high-performance electric cars, combined with its expertise in design and luxury, ensures that Jaguar will continue to shape the culture of luxury and performance for years to come. As Jaguar moves toward a fully electric future, it remains committed to delivering vehicles that provide not only exceptional performance but also the same level of luxury and sophistication that the brand has always been known for.

Chapter 16: Jaguar's Global Expansion: How the Brand Conquered the World

While Jaguar's roots are firmly planted in the United Kingdom, the brand's global success story is a testament to its universal appeal. Since the launch of its first vehicles in the 1940s, Jaguar has evolved from a niche British carmaker into one of the most renowned luxury brands in the world. Jaguar's ability to adapt to new markets, create products that resonate with diverse consumer bases, and maintain its legacy of performance and sophistication has been key to its global expansion.

Breaking Into the U.S. Market

One of Jaguar's earliest and most significant forays into the global market came in the 1950s, when the brand made its mark in the United States. As American car buyers became increasingly interested in European luxury cars, Jaguar found a welcoming market for its vehicles. The XK120, with its combination of sleek design, high performance, and relatively affordable price, became a hit in the U.S. The E-Type, which debuted in 1961, further solidified Jaguar's status as a top-tier luxury car manufacturer, not just in Britain but internationally.

Jaguar's American success was driven by its ability to appeal to the desires of affluent consumers looking for cars that were both exclusive and exhilarating. The E-Type, in particular, captured the imagination of American drivers, thanks to its iconic design, speed, and relatively lower price compared to other European sports cars. As the 1960s and 1970s saw more American consumers embracing European brands, Jaguar became a symbol of British sophistication in the U.S., with the XJ6 and XJ12 sedans becoming favorites among celebrities and business magnates.

Jaguar's Expansion into Europe and Beyond

In Europe, Jaguar's reputation grew steadily, particularly in countries like Germany, France, and Italy, where luxury and performance were highly valued. The introduction of models like the XJ Series further cemented Jaguar's place in the European luxury market. The brand's combination of innovative design and powerful performance attracted a broad range of customers, from high-ranking government officials to wealthy individuals.

While Jaguar's European success was significant, the company continued to expand into emerging markets during the late 20th and early 21st centuries. The rise of the global middle class, particularly in markets such as China and India, opened up new opportunities for Jaguar. In China, the world's largest car market, Jaguar found an enthusiastic audience for its luxury cars, with models like the XF and the XJ becoming popular choices among affluent buyers.

Jaguar also made headway in the Middle East, where its luxury sedans and sports cars found a home among the region's wealthy elite. The brand's association with British craftsmanship and its ability to offer a superior driving experience made Jaguar highly desirable in markets known for their love of luxury vehicles.

Jaguar and the Role of Globalization

Jaguar's global expansion was not just about selling cars in new markets—it was also about adapting to the cultural and economic climates of different regions. The company was quick to recognize that different markets had different needs and preferences, and it responded accordingly. For example, in markets like the U.S. and China, where consumers preferred larger vehicles, Jaguar expanded its range of luxury sedans and SUVs to cater to these demands. The introduction of models like the F-PACE SUV allowed Jaguar to compete in the rapidly growing luxury SUV segment, further boosting its global appeal.

Jaguar's strategic decisions to partner with global corporations, such as the collaboration with Tata Motors in 2008, also helped secure its global reach. Tata's ownership of Jaguar allowed the company to expand its manufacturing base and enter new markets with greater resources, particularly in emerging economies. Jaguar benefited from the synergies within Tata Motors, enabling it to improve its production processes, share technology, and gain access to new markets.

The Future of Jaguar's Global Presence

Looking to the future, Jaguar is focused on maintaining its strong global presence while adapting to the shifting dynamics of the automotive industry. As the world moves towards sustainable and electric mobility, Jaguar's commitment to producing high-performance electric vehicles like the I-PACE positions the brand to continue its global growth. The shift towards electric vehicles is a key part of Jaguar's strategy to maintain its relevance in both established and emerging markets.

Jaguar's expansion into new markets, coupled with its commitment to performance, luxury, and sustainability, ensures that the brand will remain a key player on the global stage. With the world increasingly seeking eco-friendly alternatives without sacrificing luxury or performance, Jaguar is well-positioned to thrive as a global automotive leader for years to come.

Chapter 17: Jaguar and British Luxury: The Intersection of Performance and Prestige

For decades, Jaguar has represented the epitome of British luxury, perfectly balancing refined elegance with world-class performance. The brand's unique ability to blend luxurious materials, sophisticated engineering, and exhilarating performance has made it an iconic name in the world of luxury automobiles. In this chapter, we explore how Jaguar has come to define British luxury and what sets it apart from other manufacturers in the high-end automotive market.

The Roots of British Luxury

The concept of British luxury in automobiles is deeply rooted in a tradition of craftsmanship, exclusivity, and understated elegance. British luxury cars have long been synonymous with quality, craftsmanship, and a refined aesthetic. From Rolls-Royce to Bentley, the U.K. has produced some of the world's most prestigious automobile manufacturers. Jaguar, with its unique combination of beauty, performance, and craftsmanship, has earned its place among these esteemed names.

Jaguar's approach to luxury has always been a little different. While other British brands focused on creating cars that were grand and opulent, Jaguar was driven by the idea that luxury could be combined with thrilling performance. From the early days of the brand's history, Jaguar made it clear that it was a company that wanted to provide more than just comfort and refinement—it sought to create an exciting and dynamic driving experience.

The E-Type, with its stunning design and remarkable performance, is perhaps the perfect example of how Jaguar embodies British luxury. The car's sleek lines, paired with a top speed that made it one of the fastest in the world, captured the imagination of those who sought both beauty and power. The E-Type was not simply a car; it was a symbol of the kind of luxury that was attainable, dynamic, and passionate. This blend of elegance and performance became the core of Jaguar's identity.

Modern British Luxury: The Jaguar XJ and Beyond

The XJ series of sedans has long been the flagship of Jaguar's luxury offerings, and it has become synonymous with British prestige. The XJ, with its elegant design, exceptional comfort, and impeccable craftsmanship, represents the perfect fusion of luxury and performance. The interior of the XJ is designed with the utmost attention to detail, using premium materials such as fine leather, rich wood veneer, and polished metal. The car's elegant yet understated appearance reflects Jaguar's belief that luxury should be refined, not ostentatious.

The driving experience in a Jaguar XJ is one of seamless comfort and precision. The suspension system ensures that even the roughest roads provide a smooth, quiet ride, while the advanced engineering ensures that the car remains agile and responsive when it comes to handling. The XJ is a perfect example of how Jaguar has taken the concept of British luxury and combined it with the kind of performance that makes driving a true pleasure.

The 21st century saw Jaguar continue to redefine luxury, particularly with the launch of the XF and F-Type models. The XF sedan, with its modern design, technology, and engineering, helped Jaguar expand its reach to a broader audience, while the F-Type represented the brand's return to its sports car roots. Both vehicles demonstrated Jaguar's ability to maintain its reputation for performance and elegance while also adapting to the changing tastes of the luxury car market.

Jaguar's Appeal to the Elite

What truly sets Jaguar apart from other luxury brands is its ability to appeal to the elite while remaining accessible. Unlike the ultra-exclusive Rolls-Royce or Bentley, Jaguar offers luxury without the level of exclusivity that often comes with the former two brands. Jaguar vehicles are coveted by business leaders, celebrities, and even royalty, but they also offer more attainable options for those who want to experience British luxury without breaking the bank.

Jaguar's appeal to the elite is not just about the car itself; it's also about the experience of owning and driving a Jaguar. The company's focus on exceptional service, attention to detail, and personalized options ensures that owning a Jaguar is more than just having a vehicle—it's an experience that elevates the owner's lifestyle. This commitment to the customer experience has played a crucial role in Jaguar's success as a luxury brand, allowing it to reach a broad range of consumers who want a taste of the high life.

Jaguar's luxurious designs, combined with its

commitment to high-performance engineering, have made it a cornerstone of British automotive heritage. As the world of luxury cars continues to evolve, Jaguar remains dedicated to ensuring that its cars remain symbols of style, elegance, and driving excellence.

Chapter 18: Technological Innovations: From the Monocoque to All-Wheel Drive

Jaguar has always been at the forefront of automotive technology, pioneering innovations that have helped shape the car industry for decades. The company's commitment to blending cutting-edge technology with timeless design and exceptional performance has made it a leader in automotive engineering. In this chapter, we take a look at some of Jaguar's most important technological innovations—from the D-Type's monocoque construction to the modern all-wheel-drive systems found in the F-Type—and how these advancements have influenced both Jaguar's cars and the wider automotive industry.

Monocoque Construction: Revolutionizing Car Design

One of Jaguar's most important technological contributions to the automotive world came in the 1950s with the D-Type, a car that would not only change the course of racing history but also revolutionize car design. The D-Type introduced the monocoque construction, a method of building a car's body and chassis as a single, integrated unit. This innovation made the car lighter, stronger, and more aerodynamic, improving both its performance and safety.

Monocoque construction quickly became the standard in the automotive industry, influencing the design of everything from sports cars to everyday vehicles. Jaguar's early use of this technology in the D-Type helped position the brand as a leader in both engineering and racing, and it paved the way for many of the innovations that followed in subsequent decades.

Advancements in Suspension Systems: Ensuring Comfort and Control

Jaguar has also been a leader in developing advanced suspension systems, ensuring that its cars deliver a smooth, comfortable ride without sacrificing handling performance. The company's early development of independent suspension systems, first introduced in the Jaguar XK120, was another key innovation that helped Jaguar set itself apart from other manufacturers.

In the 1960s, Jaguar took this a step further with the introduction of the XJ6. The XJ6's fully independent suspension system provided exceptional comfort and handling, helping the car achieve the refined ride and precise handling for which Jaguar is known. The introduction of electronic and air suspension systems in later models further enhanced Jaguar's ability to deliver comfort and agility, ensuring that even the heaviest sedans and sportscars could deliver a driving experience that was both exhilarating and refined.

All-Wheel Drive: Expanding Jaguar's Reach

As the demand for all-wheel-drive vehicles increased, Jaguar took steps to expand its portfolio to include these systems. All-wheel drive is a vital technology in modern luxury cars, providing better handling, increased traction, and greater safety, especially in challenging weather conditions. Jaguar's use of all-wheel drive technology has been instrumental in ensuring that the brand's cars remain competitive in the global market.

The first major foray into all-wheel-drive came with the introduction of the Jaguar XJ in the 1980s. The car's advanced all-wheel-drive system was designed to provide enhanced stability and handling, particularly in inclement weather. Today, Jaguar's all-wheel-drive systems are a standard feature on several models, including the F-Type and the XF sedan, ensuring that the brand's performance is uncompromised regardless of the conditions.

The all-wheel-drive technology found in these modern models provides a perfect example of how Jaguar is always evolving its technology to meet the needs of its customers, combining performance with practicality to create cars that are both thrilling and safe to drive.

Future Innovations: Electric Powertrains and Beyond

Looking ahead, Jaguar's commitment to innovation continues with the transition to electric vehicles. The brand's all-electric I-PACE is a prime example of how Jaguar is combining traditional engineering excellence with cutting-edge technology. The I-PACE features an advanced electric powertrain that delivers instant torque, exceptional acceleration, and a driving experience that stays true to Jaguar's heritage of performance.

In addition to electric propulsion, Jaguar is also working on innovations in battery technology, autonomous driving systems, and connected vehicle technology. As the automotive industry moves toward an electric and autonomous future, Jaguar is at the forefront, ensuring that its cars remain technologically advanced, luxurious, and high-performing.

Jaguar's legacy of technological innovation is one of the key reasons the brand remains one of the most respected names in the automotive industry. Its commitment to pushing the boundaries of what is possible, whether in suspension systems, all-wheel drive, or electric vehicles, ensures that Jaguar will continue to set the standard for performance, luxury, and engineering excellence for years to come.

Chapter 19: The Future of Jaguar: Electric Vehicles, Autonomous Cars, and the New Mobility Era

The automotive industry is undergoing a profound transformation, driven by advances in electric propulsion, autonomous technology, and new models of mobility. As one of the most storied and innovative car manufacturers in the world, Jaguar is embracing this change and positioning itself as a leader in the new automotive era. In this chapter, we explore what the future holds for Jaguar, particularly as the company transitions to electric vehicles (EVs), explores the possibilities of autonomous driving, and adapts to the changing needs of consumers in a rapidly evolving world.

Jaguar's Commitment to Electrification

Jaguar's move toward electrification is a defining moment in the brand's history. With increasing concerns about climate change, emissions regulations, and the growing demand for sustainable mobility, Jaguar has committed to becoming a fully electric car manufacturer by 2025. This bold decision marks a pivotal shift for the brand, but it also reflects Jaguar's willingness to adapt and innovate in the face of a rapidly changing automotive landscape.

The launch of the I-PACE in 2018 marked Jaguar's first major step into the world of electric vehicles, and it has since become a cornerstone of the brand's electric future. The I-PACE, with its dynamic performance, stunning design, and eco-friendly credentials, has proven that Jaguar can deliver all the excitement and luxury that customers expect from the brand while also meeting the demands of sustainability.

Jaguar's commitment to electric vehicles goes beyond just offering a few EVs in its lineup. By 2025, Jaguar plans to have a fully electric range of vehicles, including SUVs, sedans, and sports cars. These vehicles will combine Jaguar's performance heritage with cutting-edge electric technology, ensuring that the brand remains at the forefront of the luxury automotive market.

Autonomous Driving: The Next Frontier

In addition to electrification, Jaguar is exploring the future of autonomous driving. Autonomous technology has the potential to revolutionize the way people interact with their cars, offering increased convenience, safety, and efficiency. Jaguar's parent company, Tata Motors, has been actively involved in autonomous vehicle development, and Jaguar is keen to bring this technology to its cars.

While fully autonomous cars are still a few years away from widespread adoption, Jaguar is already testing and integrating advanced driver-assistance systems (ADAS) in its vehicles. Features such as adaptive cruise control, lane-keeping assist, and automatic emergency braking are already available in several Jaguar models, providing a glimpse of what the future may hold. As autonomous technology continues to evolve, Jaguar is committed to ensuring that its vehicles remain safe, efficient, and enjoyable to drive, whether behind the wheel or in self-driving mode.

The Future of Mobility: Beyond Ownership

The way consumers view car ownership is also changing, with more people opting for shared mobility solutions, ride-hailing services, and car subscriptions. Jaguar, along with its sister brand Land Rover, is exploring new models of mobility to meet these shifting consumer preferences. The brand's focus on sustainability, performance, and luxury makes it well-suited to adapt to new mobility trends, particularly in urban environments.

Jaguar's involvement in the future of mobility extends beyond just the vehicles themselves. The brand is exploring new partnerships, technologies, and services that can help redefine the automotive experience. From innovative charging solutions for electric cars to exploring the role of autonomous vehicles in shared mobility, Jaguar is actively shaping the future of transportation.

A Bold Future: Jaguar's Vision for Tomorrow

As Jaguar embarks on this transformative journey, the future of the brand is full of promise. The transition to electric vehicles, the integration of autonomous driving technology, and the exploration of new models of mobility ensure that Jaguar will remain a key player in the automotive world for decades to come. Whether it's through groundbreaking new electric models or advancements in connected and autonomous technology, Jaguar is committed to delivering exceptional vehicles that blend performance, luxury, and sustainability.

The next chapter in Jaguar's story will be defined by

innovation, sustainability, and a continued dedication to luxury and performance. With its rich heritage, strong brand identity, and forward-thinking approach, Jaguar is poised to lead the way in the new era of automotive mobility.

Chapter 20: The Enduring Appeal of Jaguar: Why the Brand Remains a Symbol of Excellence

Jaguar's ability to remain a symbol of automotive excellence over decades is a testament to the brand's unique blend of craftsmanship, performance, and innovation. Despite the ever-changing landscape of the global automotive market, Jaguar has consistently captured the imaginations of drivers, enthusiasts, and collectors around the world. Whether it's the timeless allure of the E-Type, the cutting-edge engineering of the F-Type, or the luxury of the XJ, Jaguar has built an enduring legacy that continues to resonate today. In this chapter, we explore the factors that have contributed to the lasting appeal of the Jaguar brand and how it remains relevant in a rapidly evolving automotive world.

Timeless Design: Beauty That Endures

One of the most enduring aspects of Jaguar's appeal is its design philosophy, which has remained true to its roots while evolving with the times. Jaguar's cars have always been known for their elegance, with sleek lines, curvaceous bodywork, and an overall aesthetic that blends power with grace. The E-Type, often referred to as one of the most beautiful cars ever made, set the benchmark for design excellence when it was launched in the 1960s. Its timeless beauty is still celebrated today, and it continues to serve as a muse for designers at Jaguar, influencing the brand's modern vehicles.

Jaguar's commitment to design extends to all its models, from the sedate sophistication of the XJ series to the sporty appeal of the F-Type. Even as technology advances and consumer preferences shift, Jaguar has maintained its focus on creating cars that are not just functional but also visually stunning. The brand has learned to strike the delicate balance between bold, modern designs and the classic elegance that defines British luxury, ensuring that its cars remain desirable across generations.

Jaguar's design ethos continues to influence the automotive world, with competitors often citing Jaguar's aesthetic influence. The evolution of the F-Type is a prime example of how the brand has modernized its design language while staying true to the principles of beauty, performance, and style. Even as the automotive industry moves toward electric vehicles and autonomous cars, Jaguar's commitment to design ensures that its cars remain desirable for their aesthetics as much as their performance.

The Driving Experience: Performance Meets Luxury

At the heart of Jaguar's enduring appeal is its ability to offer a driving experience that combines exhilarating performance with unparalleled luxury. The brand's cars are crafted to be more than just vehicles; they are designed to be an experience. From the moment a driver sits behind the wheel of a Jaguar, they are immersed in a world of precision handling, responsive power, and unmatched comfort.

Whether it's the agile, track-focused handling of the F-Type or the smooth, effortless glide of the XJ, Jaguar's cars are engineered to perform at the highest level. The driving experience in a Jaguar is second to none, combining world-class suspension systems, powerful engines, and cutting-edge technology to create a driving experience that is thrilling, yet refined. This marriage of performance and luxury is what sets Jaguar apart from other luxury brands, ensuring that its cars appeal not only to those seeking luxury but also to those with a passion for driving.

The F-Type, in particular, is a testament to Jaguar's ability to deliver performance without compromise. As a modern-day successor to the E-Type, the F-Type offers a driving experience that is as dynamic as it is comfortable. Its potent engines, precision handling, and aggressive styling ensure that it remains one of the most desirable sports cars on the market today.

A Legacy of Innovation: Pushing Boundaries for the Future

Jaguar's commitment to innovation is another key factor in its enduring appeal. The brand has consistently pushed the boundaries of automotive engineering, introducing technologies that have set new standards in the industry. From the monocoque construction of the D-Type to the lightweight aluminum architecture of the F-Type, Jaguar has led the way in creating cars that are not only beautiful and fast but also technologically advanced.

In recent years, Jaguar's commitment to innovation has been showcased in its electric vehicle offerings. The I-PACE, the brand's first fully electric car, received critical acclaim for its performance, design, and sustainability, solidifying Jaguar's position as a leader in the future of automotive mobility. The company's plans for a fully electric future, set to take place by 2025, show that Jaguar remains at the cutting edge of automotive technology.

As the automotive world moves towards sustainability, electric mobility, and autonomous driving, Jaguar is positioning itself to lead the way. The brand's ability to innovate and adapt ensures that it will remain relevant in the ever-changing automotive landscape, while continuing to deliver the luxury, performance, and style that have defined it for decades.

The Allure of Jaguar: A Brand for the Discerning Driver

Jaguar has always been a brand for those who seek something more from their vehicles. The brand appeals to drivers who want to experience the thrill of the road without sacrificing comfort or elegance. It attracts those who appreciate the finer things in life—luxury, performance, and design—and those who recognize the heritage and legacy of a brand that has shaped the automotive world.

As the world continues to change and the automotive industry evolves, Jaguar's ability to stay true to its core values of performance, luxury, and design will ensure that its allure remains as strong as ever. Jaguar has built an empire not just on the quality of its cars but also on the emotions they evoke. For those who appreciate the art of driving, Jaguar will always remain a symbol of excellence.

Chapter 21: Jaguar in Popular Culture: A Symbol of Glamour and Speed

Jaguar's cars have long been a fixture in popular culture, representing both the glamour of luxury and the thrill of performance. Whether it's a sleek E-Type in a 1960s film or a powerful F-Type on the racetrack, Jaguar has become synonymous with style, sophistication, and speed. In this chapter, we explore how Jaguar's vehicles have been featured in films, television, music, and other forms of popular culture, and how the brand has maintained its status as a symbol of excellence, luxury, and excitement.

The E-Type: A Hollywood Icon

Perhaps no Jaguar model has had as profound an impact on popular culture as the E-Type. Launched in 1961, the E-Type quickly became a symbol of cool, sophistication, and youthful rebellion. Its stunning design and exhilarating performance made it a favorite of celebrities, and it wasn't long before the E-Type became a Hollywood icon. The E-Type made appearances in countless films, often used as a symbol of status and sophistication. In the 1960s, its sleek lines and aggressive stance became synonymous with the cultural zeitgeist of the time.

One of the most famous appearances of the E-Type in popular culture was in the 1966 film *The Graduate*, where Dustin Hoffman's character, Benjamin Braddock, drives an E-Type while attempting to seduce the older Mrs. Robinson. The car's appearance in the film solidified its status as a symbol of allure, sophistication, and the spirit of the 1960s. The E-Type was also featured in numerous other films, including *Austin Powers: International Man of Mystery*, where it was humorously used as a time machine for the quirky character Dr. Evil.

The E-Type's influence extended beyond the silver screen; it became a symbol of glamour and performance in the real world as well. Countless celebrities—from actors and musicians to business moguls—chose the E-Type as their car of choice. Its design, combined with its speed and accessibility, made it a symbol of success, freedom, and style.

Jaguar in the Bond Franchise: A Reflection of Elegance and Power

Jaguar's association with popular culture continued in the 1990s and 2000s, particularly with its presence in the

James Bond franchise. While Bond is traditionally associated with Aston Martin, Jaguar vehicles have also made their mark in the series. Notably, in *Die Another Day* (2002), Pierce Brosnan's Bond drove a Jaguar XKR in a dramatic chase sequence, solidifying the car's status as a symbol of sophistication and power. The sleek, aggressive design of the XKR matched the dynamic, action-packed world of Bond, making it the perfect choice for a film where style and performance are key.

In subsequent Bond films, Jaguar continued to feature as the go-to car for luxurious, high-performance driving. Jaguar's association with 007 cemented its place in the cultural lexicon as a brand that exudes power, grace, and speed—characteristics that align perfectly with the image of the iconic spy.

The F-Type: Modern Glamour and the New Jaguar

As Jaguar entered the 21st century, the brand embraced a new era of performance, and its cars continued to make appearances in popular culture. The F-Type, launched in 2013, quickly became the modern face of Jaguar, embodying the same values of speed, beauty, and sophistication that the brand has been known for since its inception. The F-Type has appeared in films, television shows, and advertisements, continuing the tradition of Jaguar as a symbol of glamour and excitement.

In films such as *Spooks: The Greater Good* (2015) and *The Transporter Refueled* (2015), the F-Type was portrayed as a car for the elite, used by characters who were wealthy, stylish, and adventurous. The F-Type has also been used in advertising campaigns that highlight its design and performance, showcasing the brand's commitment to creating cars that are as exciting to drive as they are to look at.

Jaguar in Music and Fashion

Jaguar's influence is not confined to the silver screen. The brand has also played an important role in music and fashion, frequently appearing in music videos, photo shoots, and fashion shows. From musicians like The Rolling Stones, who have been seen driving E-Types, to fashion icons such as Kate Moss, who has posed with Jaguar vehicles for high-profile shoots, the brand has become synonymous with style and luxury.

Jaguar's cars are often chosen for their visual appeal and the statement they make about the owner. Whether it's the F-Type, with its modern lines and cutting-edge technology, or the timeless beauty of the E-Type, Jaguar's vehicles continue to be a symbol of status, sophistication, and style.

A Lasting Cultural Legacy

Jaguar's presence in popular culture reflects the brand's status as a symbol of luxury, speed, and elegance. From its iconic E-Type to its modern F-Type, Jaguar's vehicles have become cultural touchstones, featured in films, music, and fashion. As the automotive industry continues to evolve, Jaguar remains committed to maintaining its place in the cultural landscape, creating cars that are not only technologically advanced but also timelessly stylish.

Jaguar's legacy in popular culture ensures that the brand will remain a symbol of excellence for generations to come, a brand that continues to embody the intersection of glamour and speed.

Chapter 22: The Future of Jaguar: Challenges and Opportunities in the Electric Age

As the automotive industry faces profound changes, Jaguar is poised to navigate the future with a renewed sense of purpose and a commitment to innovation. The shift toward electrification, the integration of autonomous technologies, and changing consumer preferences all present new challenges for Jaguar, but they also offer exciting opportunities for the brand to redefine its legacy for the 21st century. In this chapter, we explore the challenges and opportunities that Jaguar will face as it enters the new electric age, and how the company plans to continue its tradition of excellence in the face of evolving industry demands.

The Push for Electrification

The future of Jaguar lies in its transition to fully electric vehicles, as the company has committed to becoming a 100% electric brand by 2025. This bold shift is not only a response to growing concerns about climate change and emissions but also a recognition of the changing preferences of consumers who are increasingly seeking sustainable alternatives to traditional gasoline-powered vehicles. Jaguar's move toward electrification is part of a broader strategy that aims to make the brand a leader in luxury electric mobility.

Jaguar's first significant step into the electric vehicle (EV) market came with the launch of the I-PACE in 2018. The I-PACE has received widespread acclaim for its dynamic performance, innovative design, and sustainable technology, proving that Jaguar can create electric cars that maintain the performance and luxury for which the brand is known. The I-PACE is not just an electric SUV; it is a statement about Jaguar's commitment to the future of mobility, offering a driving experience that is both thrilling and environmentally responsible.

With the I-PACE as the foundation, Jaguar plans to expand its electric vehicle lineup, introducing new models that offer the same exciting performance, cutting-edge technology, and elegant design. The shift to electric vehicles will involve significant changes in the manufacturing process, from the development of new electric powertrains to the integration of lightweight materials and the enhancement of battery technology. Jaguar's focus on innovation will ensure that it remains at the forefront of the electric revolution, creating vehicles that offer both sustainability and the thrilling performance that the brand is known for.

Autonomous Technology: Shaping the Future of Mobility

In addition to electrification, autonomous driving technology presents another major opportunity for Jaguar. As the automotive world moves toward self-driving vehicles, Jaguar is investing in the research and development of autonomous systems to integrate into its cars. Autonomous driving promises to revolutionize the way we experience transportation, offering new levels of safety, convenience, and efficiency.

Jaguar's parent company, Tata Motors, has been working on autonomous technologies, and Jaguar is positioning itself to benefit from these advancements. While fully autonomous vehicles are still some years away from widespread adoption, Jaguar has already begun to integrate driver-assistance features into its cars. Features such as adaptive cruise control, lane-keeping assist, and automated emergency braking are available in many of the brand's models, showcasing Jaguar's commitment to enhancing safety and convenience for its customers.

The development of autonomous vehicles will also offer Jaguar new opportunities to innovate in the realm of design and user experience. As self-driving technology evolves, Jaguar will be able to create cars that offer more than just transportation—they will become mobile living spaces, equipped with advanced connectivity features and a focus on comfort and relaxation. By embracing autonomous technology, Jaguar has the potential to transform the driving experience in ways that are both exciting and practical.

Navigating the Challenges of Sustainability

As Jaguar moves into the future, sustainability will be at the heart of its strategy. The environmental impact of the automotive industry has come under increasing scrutiny in recent years, and Jaguar has committed to reducing its carbon footprint and embracing environmentally friendly practices in all aspects of its operations. This includes not only the production of electric vehicles but also the use of sustainable materials, energy-efficient manufacturing processes, and efforts to reduce waste.

Jaguar's push for sustainability will require ongoing innovation and adaptation, as the company explores new ways to create cars that are both environmentally responsible and luxurious. From using recycled materials in the interior of vehicles to exploring new battery technologies that are more energy-efficient and have a lower environmental impact, Jaguar is committed to playing its part in building a more sustainable future.

The Road Ahead: Embracing Innovation and Excellence

The future of Jaguar is one that embraces innovation while staying true to the values that have made the brand an icon of luxury, performance, and design. Electrification, autonomous driving, and sustainability present both challenges and opportunities for Jaguar, but the company's legacy of excellence in engineering and design positions it to thrive in the changing automotive landscape.

Jaguar's commitment to pushing the boundaries of what is possible will ensure that it remains a leader in the luxury car market. Whether it's through the development of new electric vehicles, the integration of cutting-edge technology, or the creation of vehicles that redefine the driving experience, Jaguar's future is filled with promise and excitement.

As Jaguar enters this new chapter in its history, it remains dedicated to delivering cars that offer more than just performance—they offer a vision of the future of mobility, where luxury, sustainability, and innovation come together in perfect harmony.

Chapter 23: Conclusion: The Jaguar Legacy – A Brand That Transcends Time

As we reach the conclusion of this journey through Jaguar's rich history, it is clear that the brand's legacy is one that transcends time. From its humble beginnings as a sidecar manufacturer to its rise as a global leader in luxury and performance automobiles, Jaguar has consistently set the standard for excellence in the automotive world. The company's commitment to craftsmanship, innovation, and design has ensured that its cars continue to captivate enthusiasts and collectors around the world.

A Brand Built on Innovation and Excellence

At the heart of Jaguar's success is its ability to innovate. From the introduction of the monocoque construction in the D-Type to the development of the electric I-PACE, Jaguar has never been content to rest on its laurels. The company's engineers and designers have always been driven by a desire to push the boundaries of what is possible, creating cars that are not only technologically advanced but also beautiful, thrilling to drive, and timeless in their appeal.

Jaguar's legacy of innovation is not just about groundbreaking technologies—it's about creating a driving experience that excites and inspires. Whether it's the exhilarating performance of the F-Type, the luxurious comfort of the XJ, or the sustainability of the I-PACE, Jaguar's cars have always been designed to offer something more than mere transportation. They offer an experience that connects the driver to the road, that evokes emotions, and that creates lasting memories.

Timeless Elegance and Design

Another defining characteristic of the Jaguar brand is its commitment to timeless design. The cars produced by Jaguar are not just machines—they are works of art. From the elegant curves of the E-Type to the modern lines of the F-Type, Jaguar's design philosophy has always been rooted in beauty, performance, and purpose. The brand has created cars that not only perform at the highest level but also stand as icons of automotive design, admired and desired by collectors and enthusiasts alike.

Jaguar's design legacy extends beyond just the exterior of its cars. The brand's focus on interior craftsmanship, with luxurious materials and attention to detail, ensures that every Jaguar is a pleasure to drive and own. The interiors of Jaguar cars are as meticulously crafted as their exteriors, offering an environment that combines comfort, style, and modern technology.

A Brand for the Future

As Jaguar looks to the future, the company remains firmly committed to its legacy of performance, luxury, and design. The shift to electric vehicles, the integration of autonomous technologies, and the ongoing pursuit of sustainability will shape the future of the brand, but Jaguar's core values will remain the same. The brand's focus on innovation, excellence, and creating exceptional driving experiences will continue to guide its development as it embraces the challenges and opportunities of the new mobility era.

Jaguar's journey is far from over, and as it transitions to an electric future, it will continue to inspire future generations of drivers. The brand's rich heritage, combined with its forward-thinking approach, ensures that Jaguar will remain a symbol of luxury, performance, and design for years to come. The legacy of Jaguar is one of ambition, creativity, and excellence—a legacy that transcends time and continues to captivate the world.

Jaguar's Enduring Impact

From its historic victories at Le Mans to its groundbreaking designs on the road, Jaguar has left an indelible mark on the automotive industry. It is a brand that has consistently embodied the finest in British engineering, blending performance with luxury in a way that few can match. Jaguar's impact on car culture, popular media, and the automotive world is profound, and its legacy will continue to shape the future of the industry.

As Jaguar moves forward into the electric age, the essence of the brand will remain intact. The company's commitment to creating cars that are not just modes of transportation but objects of desire and joy will ensure that Jaguar's legacy remains as powerful as ever.

Jaguar is a brand that transcends time—a brand that will continue to drive the future of automotive excellence.

Appendix A: Key Models and Milestones

Jaguar has produced numerous iconic cars throughout its history, each of which has played a significant role in shaping the brand's identity and its reputation as a leader in luxury and performance automobiles. This appendix provides an overview of some of the most important models in Jaguar's history and the milestones that helped define the brand's legacy.

1. Swallow Sidecar Company (1922 - 1935)

Before the Jaguar name was born, the company was known as Swallow Sidecar Company, founded by William Lyons and William Walmsley in 1922. Initially focused on manufacturing sidecars for motorcycles, the company soon began to build custom bodies for small cars, beginning with the Austin Seven. This period marked the beginning of Jaguar's dedication to fine craftsmanship and bespoke automotive design.

Milestone: In 1935, the company changed its name to Jaguar Cars Ltd, signaling the start of its shift to producing fully designed automobiles.

2. SS Jaguar 2.5 Litre (1935)

The first car to bear the "Jaguar" name was the SS Jaguar 2.5 Litre, launched in 1935. It was a sleek, powerful saloon that reflected the growing ambitions of the company. The car featured a streamlined design, and its performance was enhanced by the use of a 2.5-liter engine, a combination of elegance and speed that would become synonymous with Jaguar.

Milestone: The SS Jaguar was Jaguar's first significant step toward establishing its reputation as a manufacturer of stylish, high-performance automobiles.

3. XK120 (1948)

One of the most important milestones in Jaguar's history, the XK120 was introduced in 1948. This car was a game-changer in both design and performance. It featured an advanced 3.4-liter engine capable of reaching 120 mph, making it the world's fastest production car at the time. The XK120's stunning looks and exceptional performance helped establish Jaguar as a premier performance car manufacturer.

Milestone: The XK120's success both on the road and in motorsports, including wins at the Mille Miglia, set the stage for Jaguar's dominance in sports cars.

4. XK140 (1954)

The XK140 was an updated version of the XK120, featuring improvements in handling, comfort, and design. With a more refined interior, enhanced suspension, and upgraded engine options, the XK140 continued Jaguar's focus on blending luxury with performance.

Milestone: The XK140's success in motorsports, including continued victories at events like Le Mans, further solidified Jaguar's place in automotive history.

5. XK150 (1957)

The XK150 was the final model in the XK series, offering more refined styling, improved performance, and better handling. It was powered by a more advanced 3.4-liter engine and featured a redesigned front end, providing a more aggressive look. The XK150 also introduced disc brakes, a significant technological innovation in the automotive world.

Milestone: The introduction of disc brakes in the XK150 was a major step forward in automotive safety and technology, influencing car manufacturers worldwide.

6. The E-Type (1961)

Widely regarded as one of the most beautiful cars ever made, the Jaguar E-Type, launched in 1961, became an instant icon. With its sleek, aerodynamic design and a top speed of over 150 mph, the E-Type was an engineering marvel. It featured a 3.8-liter engine and a sophisticated monocoque body, setting new standards in both design and performance.

Milestone: The E-Type's stunning design and performance made it a cultural icon, and its success in motorsports helped elevate Jaguar's global status. Enzo Ferrari famously called it "the most beautiful car ever made."

7. XJ6 (1968)

The XJ6 was a significant milestone in Jaguar's history as it marked the company's entry into the luxury sedan market. With its elegant design, advanced technology, and high-performance capabilities, the XJ6 quickly became the flagship of Jaguar's lineup. It featured a sophisticated independent rear suspension system, offering an unrivaled driving experience in terms of comfort and handling.

Milestone: The XJ6's success helped solidify Jaguar's position as a maker of world-class luxury sedans and marked the beginning of Jaguar's dominance in the high-end automotive market.

8. XJ220 (1992)

The XJ220 was Jaguar's entry into the world of supercars. Powered by a 3.5-liter twin-turbo V6 engine, the XJ220 was capable of reaching a top speed of 217 mph, making it one of the fastest production cars in the world at the time. Despite some controversy over its development, including changes in its engine configuration and production numbers, the XJ220 remains one of the most iconic and desirable cars in Jaguar's history.

Milestone: The XJ220 showcased Jaguar's ability to compete with European supercar manufacturers and highlighted the brand's engineering prowess.

9. XK8 (1996)

The XK8 marked Jaguar's return to the luxury sports car market, featuring a modernized design that incorporated both performance and luxury. The XK8 was powered by a 4.0-liter V8 engine, offering exhilarating performance and luxury in equal measure. The model's success led to the introduction of the XKR, a supercharged version of the XK8.

Milestone: The XK8's success helped revive Jaguar's sports car lineup and set the stage for the brand's continued focus on performance-oriented vehicles in the 21st century.

10. XF (2008)

The Jaguar XF was a game-changing model for the brand, featuring a sleek design and advanced technology that helped redefine Jaguar's presence in the luxury sedan market. It featured improved handling, a more refined interior, and a range of powerful engine options, combining performance with luxury in a way that appealed to a new generation of buyers.

Milestone: The launch of the XF in 2008 represented a major shift for Jaguar, moving toward a more modern and innovative design language that would influence future models like the XE and F-Type.

11. I-PACE (2018)

The I-PACE is Jaguar's first fully electric vehicle and represents the company's commitment to a sustainable future. With an all-electric powertrain, the I-PACE offers exceptional performance, including rapid acceleration and impressive handling, while maintaining Jaguar's signature luxury and comfort. It has received numerous accolades, including the 2019 World Car of the Year award.

Milestone: The I-PACE marks Jaguar's entry into the electric vehicle market, solidifying its position as a leader in the transition to sustainable mobility. The model represents Jaguar's bold move toward an all-electric future.

Summary of Key Milestones

Jaguar has come a long way since its early days as the Swallow Sidecar Company, and these key models and milestones highlight the brand's commitment to design excellence, performance, and innovation. From the iconic XK120 to the revolutionary I-PACE, Jaguar has consistently pushed the boundaries of what is possible in automotive design and technology. The company's rich legacy is one of resilience, passion, and a constant drive to create cars that are both beautiful and high-performing.

Jaguar's ability to evolve with the times while maintaining its focus on luxury, performance, and innovation ensures that the brand will continue to lead the automotive industry into the future.

Appendix B: Jaguar's Impact on Automotive Design

Jaguar has long been recognized for its groundbreaking contributions to automotive design, setting the standard for both aesthetic beauty and engineering innovation. The brand has been instrumental in shaping the automotive world, not only by creating visually striking cars but by introducing new design philosophies and technologies that have influenced the broader industry. In this appendix, we take a look at some of Jaguar's most important contributions to automotive design and how these innovations have impacted both the brand and the wider car industry.

1. The Monocoque Construction: A Revolution in Car Design

One of Jaguar's most important design innovations was the introduction of the monocoque construction in the 1950s. Originally developed for aircraft, this design technique was applied to the D-Type race car, which debuted in 1954. The monocoque construction integrated the body and chassis into a single, unified structure, making the car both lighter and stronger than traditional body-on-frame designs.

This innovation proved to be revolutionary, allowing Jaguar to achieve greater speed, agility, and handling on the racetrack. The D-Type's use of the monocoque design was a critical factor in its success at the 24 Hours of Le Mans, where it won three consecutive times from 1955 to 1957. The concept quickly spread throughout the automotive industry, and today, the monocoque design is the standard for most modern vehicles, particularly in racing and high-performance cars.

Jaguar's pioneering use of this construction technique set the stage for the development of lighter, more efficient, and safer vehicles across the entire industry, fundamentally changing car design for generations to come.

2. The E-Type: A Design Icon for the Ages

The Jaguar E-Type, launched in 1961, is widely regarded as one of the most beautiful cars ever made. Designed by Malcolm Sayer, the E-Type's aerodynamic shape, long bonnet, and sleek curves were a significant departure from the more angular designs of the era. It was an instant sensation, attracting attention from car enthusiasts, celebrities, and even automotive experts like Enzo Ferrari, who called it "the most beautiful car ever made."

The E-Type was more than just a pretty face—it was a technological marvel that combined beauty with performance. The car featured advanced engineering, including independent suspension and disc brakes, setting a new standard for road cars. Its performance capabilities, including a top speed of over 150 mph, made it one of the fastest production cars of its time.

The E-Type's design had a profound impact on the automotive industry, influencing countless car manufacturers and helping to establish Jaguar as a leader in both design and performance. Its timeless beauty continues to inspire designers today, and it remains a symbol of the golden age of motoring.

3. The XJ Series: Defining Luxury Sedan Design

The Jaguar XJ, first introduced in 1968, set a new benchmark for luxury sedan design. Designed by Sir William Lyons, the XJ combined the best elements of Jaguar's design philosophy—elegance, performance, and comfort—into one cohesive package. The XJ's refined, elegant lines made it a standout in the luxury car market, and its use of cutting-edge technology, including independent rear suspension, helped elevate its performance to new heights.

Jaguar's focus on craftsmanship and luxury was evident in every detail of the XJ, from its sumptuous leather interiors to its high-quality materials and finishes. The XJ quickly became the car of choice for business leaders, celebrities, and even members of the British royal family, further cementing Jaguar's status as a manufacturer of luxury vehicles.

The XJ's design influenced a generation of luxury sedans, and its lasting impact can still be seen in the designs of modern executive cars. The XJ's blend of sophisticated design and high-performance engineering helped Jaguar establish a reputation for creating cars that were as luxurious as they were exhilarating to drive.

4. Lightweight Materials: Advancing Performance and Sustainability

Jaguar has long been at the forefront of using lightweight materials in car design, seeking to improve both performance and fuel efficiency. The brand's focus on lightweight construction can be traced back to the

1950s, when the D-Type race car featured a lightweight aluminum body that helped the car achieve faster speeds and improved handling.

This commitment to lightweight design continued with models like the XK120 and the E-Type, both of which used lightweight materials to enhance performance. In more recent years, Jaguar has continued to innovate in the use of lightweight materials, incorporating advanced aluminum architecture in models such as the F-Type and the latest-generation XJ. This approach reduces weight, improves fuel efficiency, and enhances the car's handling, allowing Jaguar to continue its legacy of performance and innovation.

Jaguar has also embraced sustainability by using recycled aluminum in its production processes, helping to reduce the environmental impact of manufacturing. The brand's focus on lightweight materials not only improves the driving experience but also ensures that Jaguar remains at the cutting edge of automotive technology.

5. The Evolution of the Sports Car: The F-Type

Jaguar's return to the sports car market in the 21st

century with the F-Type demonstrated the brand's commitment to staying relevant in a changing industry. Launched in 2013, the F-Type combined the sleek, muscular design of Jaguar's past sports cars with modern performance and technology. The F-Type's design was influenced by the E-Type, with its long hood, short rear deck, and wide stance. The car's aggressive lines and sculpted bodywork gave it a distinctive, dynamic look that set it apart from other sports cars.

Underneath the F-Type's beautiful exterior lies a powerful engine lineup, ranging from the supercharged V6 to the 5.0-liter supercharged V8, ensuring that the car delivers thrilling performance on the road. The F-Type also introduced modern technologies, including advanced suspension systems, an active exhaust, and an intuitive infotainment system, all of which combined to make it a driver-focused sports car that maintains Jaguar's tradition of innovation.

The F-Type is a prime example of how Jaguar has managed to combine traditional design elements with modern technologies, ensuring that its cars remain relevant and exciting in an ever-evolving market. The F-Type has become a modern icon, drawing on Jaguar's rich sports car heritage while embracing the latest advancements in automotive design.

6. The I-PACE: A New Era of Electric Design

Jaguar's first fully electric vehicle, the I-PACE, represents the brand's bold leap into the future of automotive design. Launched in 2018, the I-PACE combines Jaguar's legacy of performance and design with the latest advancements in electric vehicle technology. The I-PACE's sleek, futuristic design is a departure from traditional electric vehicles, with its aerodynamic lines, bold grille, and signature LED headlights. The car's dynamic proportions and aggressive stance give it the presence of a performance SUV, while its luxurious interior offers the same level of craftsmanship that Jaguar is known for.

Under the hood, the I-PACE features a 90-kWh battery pack that powers dual electric motors, delivering 394 horsepower and 512 lb-ft of torque. The I-PACE accelerates from 0 to 60 mph in just 4.5 seconds, proving that electric vehicles can deliver the same exhilarating performance as traditional combustion engines.

The I-PACE is a significant milestone for Jaguar, as it marks the brand's commitment to sustainability without sacrificing performance or luxury. The car's design showcases Jaguar's ability to innovate while remaining true to the brand's core values, and it represents the future of Jaguar as the company moves toward an all-electric lineup.

7. Jaguar's Design Philosophy: A Lasting Legacy

Throughout its history, Jaguar has maintained a design philosophy that emphasizes beauty, performance, and innovation. Whether it's the graceful curves of the E-Type or the modern, sleek lines of the F-Type, Jaguar's design ethos has always been about creating cars that are both visually stunning and engineered to perform at the highest level. The brand's commitment to craftsmanship and attention to detail ensures that every Jaguar car is a work of art, built to stand the test of time.

Jaguar's influence on automotive design extends beyond its own cars. The brand's innovative use of lightweight materials, its emphasis on aerodynamics, and its focus on creating a dynamic driving experience have set trends that other manufacturers have followed. Jaguar's design legacy is one that will continue to shape the automotive world for years to come, and its cars will remain icons of beauty and performance for generations.

Summary

Jaguar's contributions to automotive design have been profound and lasting. From the revolutionary monocoque construction of the D-Type to the modern elegance of the I-PACE, Jaguar has consistently pushed the boundaries of what is possible in automotive engineering. The brand's focus on performance, luxury, and innovative design continues to set it apart as one of the most respected and influential car manufacturers in the world. Jaguar's impact on automotive design is undeniable, and its legacy will continue to shape the future of the industry for years to come.

Appendix C: Jaguar's Motorsports Legacy

Jaguar's presence in motorsport has been integral to its identity, helping the brand achieve engineering excellence, sharpen its technological capabilities, and build a reputation as a leader in performance. The company's involvement in motorsport has not only contributed to its success on the racetrack but has also informed the design and development of its road cars, ensuring they remain at the cutting edge of automotive innovation. In this appendix, we explore Jaguar's long and storied motorsports legacy, highlighting key achievements, race victories, and technological innovations that have shaped the brand's success both on and off the track.

1. The Early Years: Racing Roots and the D-Type

Jaguar's commitment to motorsport began almost as soon as the brand was established. Early models like the XK120 competed in various events, but it was the introduction of the D-Type in 1954 that truly set the brand apart in motorsports. Designed specifically for endurance racing, the D-Type became one of Jaguar's most iconic race cars, and its successes at the 24 Hours of Le Mans helped cement Jaguar's reputation as a performance-oriented brand.

The D-Type's success was a result of several innovations in design, including its lightweight monocoque construction, which made it more aerodynamically efficient and easier to handle during high-speed races. This design was a significant leap forward in automotive engineering and would go on to influence the design of many future race cars. The D-Type secured three consecutive victories at the prestigious Le Mans race in 1955, 1956, and 1957, establishing Jaguar's dominance in endurance racing.

Milestone: The D-Type's back-to-back victories at Le Mans from 1955 to 1957 highlighted Jaguar's commitment to innovation and solidified its place in the world of motorsport.

2. The E-Type and its Racing Influence

While the Jaguar E-Type is primarily celebrated for its stunning design and street performance, its connection to motorsport is also significant. Launched in 1961, the E-Type was not only one of the most beautiful sports cars of its time, but it also proved itself capable in racing. Privateer racers, who used the E-Type in various events, found success in both track and road racing, further elevating the car's iconic status.

Jaguar's racing success with the E-Type, particularly in European GT racing, reinforced the brand's reputation for combining beauty with performance. The car's aerodynamic shape, power, and nimble handling made it a formidable competitor in racing events, continuing Jaguar's legacy of high-performance sports cars.

Milestone: The E-Type's involvement in motorsport, even though it was not a factory-supported effort, showed how Jaguar's street cars were engineered for performance and set the stage for the brand's motorsport heritage in the 1960s.

3. The 1960s and 1970s: Le Mans and the Jaguar XJ

Jaguar's participation in Le Mans continued throughout the 1960s and into the 1970s, with varying levels of success. The brand shifted its focus from the D-Type to models like the Jaguar XJ, and private racing teams utilized Jaguar's vehicles in the famous endurance race. Though Jaguar didn't win as frequently in this period, the brand remained a significant presence on the racetrack.

During this era, Jaguar introduced the XJ13, a prototype developed for Le Mans. Although the car never raced in a major event due to the brand's withdrawal from factory-supported racing efforts, it became a legend in its own right. The XJ13 is remembered for its innovative design, with a 5.0-liter V12 engine and a stunning aerodynamic shape that foreshadowed the brand's focus on performance and future innovation.

Milestone: The introduction of the XJ13, while never raced in an official event, showcased Jaguar's forward-thinking approach to motorsport design and influenced future models.

4. The 1980s: The Return to Racing with the XJ-S

After a period of relative inactivity in motorsport, Jaguar returned to racing in the 1980s with the XJ-S. The XJ-S was originally designed as a luxury grand tourer but was adapted for racing and found success in various touring car events. Its performance capabilities were honed for track conditions, and it became a staple in the British Touring Car Championship (BTCC) as well as in European events.

Jaguar's performance with the XJ-S marked the

company's reentry into serious motorsport after a few decades. The car was adapted for both racing and endurance events, and its strong performance in various competitions reignited Jaguar's passion for motorsport. The XJ-S' success in this period served as a reminder that Jaguar's commitment to racing was still very much alive, and the brand was dedicated to building high-performance vehicles suited to the track.

Milestone: The XJ-S' performance in the BTCC and other touring car events marked Jaguar's return to motorsport after an extended hiatus and reignited the brand's racing pedigree.

5. Formula 1: The 2000s and Jaguar Racing

Jaguar's involvement in Formula 1 came in 2000 when the brand acquired the Stewart Grand Prix team. Rebranded as Jaguar Racing, the team's entry into F1 was seen as a bold move for the company, and it sought to compete at the highest level of motorsport. While Jaguar Racing did not achieve the level of success it hoped for, the team's presence in F1 gave the brand valuable insights into cutting-edge automotive technology and performance.

Jaguar Racing faced significant challenges, with the team struggling to keep up with the established F1 powerhouses like Ferrari, McLaren, and Williams. Despite this, Jaguar's time in F1 served as a proving ground for the development of advanced engineering solutions, many of which would eventually be integrated into Jaguar's road cars. In 2004, the team was sold to Red Bull Racing, but Jaguar's F1 journey left a lasting legacy in terms of technological development and brand exposure.

Milestone: Jaguar Racing's time in Formula 1 helped the company refine its engineering capabilities and bring cutting-edge technologies to its production vehicles, even if it didn't yield immediate on-track success.

6. The Modern Era: The F-Type and Return to Track-Ready Performance

Jaguar's most recent motorsports involvement has centered around the development of track-oriented cars and participating in racing series like the British GT Championship. The F-Type, launched in 2013, has been the company's most important sports car in decades, and its racing pedigree continues to grow as Jaguar has entered into the GT4 category of racing. The F-Type's combination of power, agility, and cutting-edge technology has made it a formidable competitor in racing events around the world.

Jaguar's recent focus on electric and hybrid technology has also led the brand to participate in Formula E, the all-electric racing series. Jaguar's involvement in Formula E is part of its strategy to embrace sustainability and showcase the performance capabilities of electric vehicles. The Jaguar I-TYPE, the brand's electric race car, has proven competitive in the series, underscoring Jaguar's commitment to performance in the electric age.

Milestone: The F-Type's success in GT racing and Jaguar's participation in Formula E highlight the brand's continued focus on high-performance, electrified motorsport.

7. The Technological Legacy of Motorsport

Jaguar's involvement in motorsport has always been about more than just winning races; it has been a means to push the boundaries of automotive technology and engineering. Many of the technologies developed for Jaguar's race cars have found their way into its production vehicles, improving performance, safety, and efficiency. From advanced suspension systems and lightweight materials to cutting-edge aerodynamics and electric powertrains, Jaguar's motorsports heritage continues to shape the future of the brand.

Milestone: Technologies like lightweight aluminum, advanced suspension systems, and aerodynamics first developed for Jaguar's race cars continue to influence the design of the company's road-going vehicles, ensuring that motorsport remains at the heart of Jaguar's engineering philosophy.

Summary

Jaguar's motorsport legacy has been a key factor in the company's success, helping the brand develop the performance, innovation, and design philosophy that defines it today. From the victories of the D-Type at Le Mans to the modern-day F-Type's track successes, Jaguar's involvement in racing has not only showcased the brand's engineering excellence but also paved the way for groundbreaking advancements in road car technology. Whether it's in endurance racing, touring car events, or electric vehicle racing, Jaguar's motorsport legacy continues to inspire and influence its approach to building world-class, high-performance cars.

Jaguar's racing history is a testament to the brand's pursuit of excellence, and as it moves toward a future of electrification and sustainability, its motorsport legacy will undoubtedly continue to shape its development in exciting and innovative ways.

About the Author

Etienne Psaila, an accomplished author with over two decades of experience, has mastered the art of weaving words across various genres. His journey in the literary world has been marked by a diverse array of publications, demonstrating not only his versatility but also his deep understanding of different thematic landscapes. However, it's in the realm of automotive literature that Etienne truly combines his passions, seamlessly blending his enthusiasm for cars with his innate storytelling abilities.

Specializing in automotive and motorcycle books, Etienne brings to life the world of automobiles through his eloquent prose and an array of stunning, high-quality color photographs. His works are a tribute to the industry, capturing its evolution, technological advancements, and the sheer beauty of vehicles in a manner that is both informative and visually captivating.

A proud alumnus of the University of Malta, Etienne's academic background lays a solid foundation for his meticulous research and factual accuracy. His education has not only enriched his writing but has also fueled his career as a dedicated teacher. In the classroom, just as in his writing, Etienne strives to inspire, inform, and ignite a passion for learning.

As a teacher, Etienne harnesses his experience in writing to engage and educate, bringing the same level of dedication and excellence to his students as he does to his readers. His dual role as an educator and author makes him uniquely positioned to understand and convey complex concepts with clarity and ease, whether in the classroom or through the pages of his books.

Through his literary works, Etienne Psaila continues to leave an indelible mark on the world of automotive literature, captivating car enthusiasts and readers alike with his insightful perspectives and compelling narratives.

Visit www.etiennepsaila.com for more.

www.ingramcontent.com/pod-product-compliance
Ingram Content Group UK Ltd.
Pitfield, Milton Keynes, MK11 3LW, UK
UKHW021919071225
9426UKWH00023B/676